SUPER MATH TRICKS

by Dr. Zondra Knapp
"Dr. Zee"
illustrated by Neal Yamamoto

Reviewed and endorsed by Tara Saraye, math teacher at Jordan Elementary School in Orange, California, and member of California Science Project

Lowell House Juvenile
Los Angeles

CONTEMPORARY BOOKS
Chicago

To Drea and David, the real Super Sleuths
–Z.K.

Publisher: Jack Artenstein
Director of Publishing Services: Rena Copperman
Executive Managing Editor: Brenda Pope-Ostrow
Editor in Chief: Amy Downing
Project Editor: Christy Tomita Maganzini
Text Design: Carolyn Wendt
Art Director: Lisa-Theresa Lenthall

Manufactured in the United States of America

ISBN:1-56565-269-X
Library of Congress Catalog Card Number: 95-18687

10 9 8 7 6 5

Lowell House books can be purchased at special discounts when ordered in bulk for premiums and special sales. Contact Department TC at the following address:

Lowell House Juvenile
2020 Avenue of the Stars
Suite 300
Los Angeles, CA 90067

CONTENTS

WELCOME TO THE SUPER SLEUTH ACADEMY

Are you ready to take the Super Sleuth challenge? On the following pages, you'll find fun and stimulating tricks, games, and puzzles, and you'll meet some super-sleuthy detectives who use math as an exciting adventure!

The book is divided into three chapters, each of which gets progressively harder. At the end of each chapter, you'll take a quiz to see how well you understood that section. Once you've completed the book, you'll get a certificate of completion and earn a spot as a Super Sleuth!

Every case begins by setting the scene, then moves into the math problem. Some of the math operations used to solve the problems in this book are:

- Cryptogram decoding: solving codes through basic addition and subtraction problems
- Sequencing: finding patterns within a set
- Ratios: a comparison of two similar things like 10:10, or 2:1
- Pattern recognition: finding a certain design or arrangement
- Line graphing: a diagram that uses a line to show changes in the value of a quantity
- Topology: the study of surfaces and shapes

- Basic geometry: lines and shapes in space; some geometric shapes are circles, squares, and triangles
- Line paradoxes: moving a line with the result that seems unbelievable but is actually true
- Tangrams: a puzzle made by cutting any shape into other shapes and forming different figures and designs

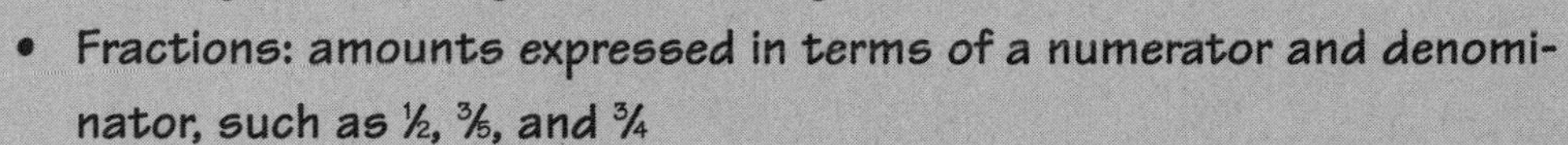

- Fractions: amounts expressed in terms of a numerator and denominator, such as $\frac{1}{2}$, $\frac{3}{5}$, and $\frac{3}{4}$
- Palindromes: numbers that read the same forward and backward
- Symmetry: the same arrangement on either side of an object that's been split down the middle

Your Super Sleuth Tools

You will need a writing journal when you solve many of the problems in your Super Sleuth investigations. You can use any one of the following suggestions for your Super Sleuth Journal:

- Three-ring notebook binder
- Spiral notebook
- Artist's notebook (a small book without any lines) and a ruler

You will also need a few good pencils with big erasers (even the best sleuths make mistakes!). Other tools needed for each trick or puzzle are listed under "Tools You'll Need."

Once you get your tools together, you'll be all set to join the academy. Put on your detective hat, sharpen your pencils, and get ready for fun. Happy sleuthing!

1: Math Sleuth Rookie

GET THE FACTS

MATHEMATICAL OPERATION: Math trick

THE SCENE: Interrogation room, police headquarters. Super Sleuth rookies, including yourself, are questioning suspect Michael Liezalot.

TOOLS YOU'LL NEED: Super Sleuth Journal, pencil

You and other Super Sleuth rookies need Mr. Liezalot's telephone number and age to get an accurate identification of this suspicious fellow, but he won't cooperate. "Tell you what," suggests one sleuth. "Let's take a break. I've got a great math trick I bet you'll find interesting." The suspect smirks and says, "Sure, why not? I'm good with numbers."

As the sleuth gives the following step-by-step instructions, you and your fellow rookies make notes in your journals and work the math trick yourselves.

Step 1: Think of your telephone number (don't include the area code), and multiply it by 2.
Step 2: Add 5 and multiply the answer by 50.
Step 3: Add your age and the number of days in one year.
Step 4: Subtract 615 from the answer, and write it down.

The suspect does all the calculations in his head, then scribbles his answer on the pad of paper and pushes it across the table.

"Can I go now?" he growls.

"In a minute," says the senior sleuth. "I just need to check this number with my partner." He hands the pad of paper to another young detective, who disappears for a moment. When he returns, he slaps a pair of handcuffs on the crook. "You're under arrest. According to our records, Mr. Liezalot, you're wanted for crimes in seven states."

As you and other rookie sleuths look on, the criminal is led away, bawling his head off like a big baby. When Mr. Liezalot finished the mathematical computations, the investigators uncovered both his phone number and his age. Once these numbers were run through a Super Sleuth database, his criminal record popped right up.

Take a look at the answer you got in your journal. The first group of numbers on the left make up your telephone number; the last number (or numbers) on the right reflects your age.

BADGE ME

MATHEMATICAL OPERATION: Cryptogram decoding

THE SCENE: Super Sleuth Academy. You and some rookies are learning to decode secret messages.

TOOLS YOU'LL NEED: Super Sleuth Journal, pencil

The Super Sleuth Academy is where junior sleuths learn the basics of good detective work. Every junior sleuth is required to take an advanced course in cryptography, the study of secret codes. Your first course assignment is to decode the secret message engraved on the official Super Sleuth badge.

You and your partner are on the brink of finally cracking the code. Both of you suspect that it is an alphanumeric code—one in which each letter of the alphabet corresponds to a certain number. You've worked out a series of subtraction problems that you feel correspond to certain letters. The next step is to solve the problems and match the numeric answers to letters of the alphabet. For example, in the first subtraction problem, 34 - 14 = 20, the number 20 is the code number for the letter A. First, copy these problems in your Super Sleuth Journal and solve them to create a key.

Key

34 - 14 = 20 = A	47 - 22 = ___ = G	59 - 47 = ___ = R
65 - 34 = ___ = B	41 - 11 = ___ = H	68 - 44 = ___ = S
80 - 40 = ___ = C	43 - 8 = ___ = I	44 - 8 = ___ = T
57 - 6 = ___ = D	77 - 32 = ___ = L	60 - 12 = ___ = U
73 - 13 = ___ = E	73 - 18 = ___ = N	100 - 9 = ___ = V
81 - 11 = ___ = F	62 - 12 = ___ = O	90 - 8 = ___ = Y

Congratulations on becoming a crackerjack code buster! Now you can use your key to decode the mysterious message on the Super Sleuth badge. Copy this sequence of numbers from the badge. Then use the key to help you match the code number with the letter of the alphabet. Finally, write the secret message in your Super Sleuth Journal.

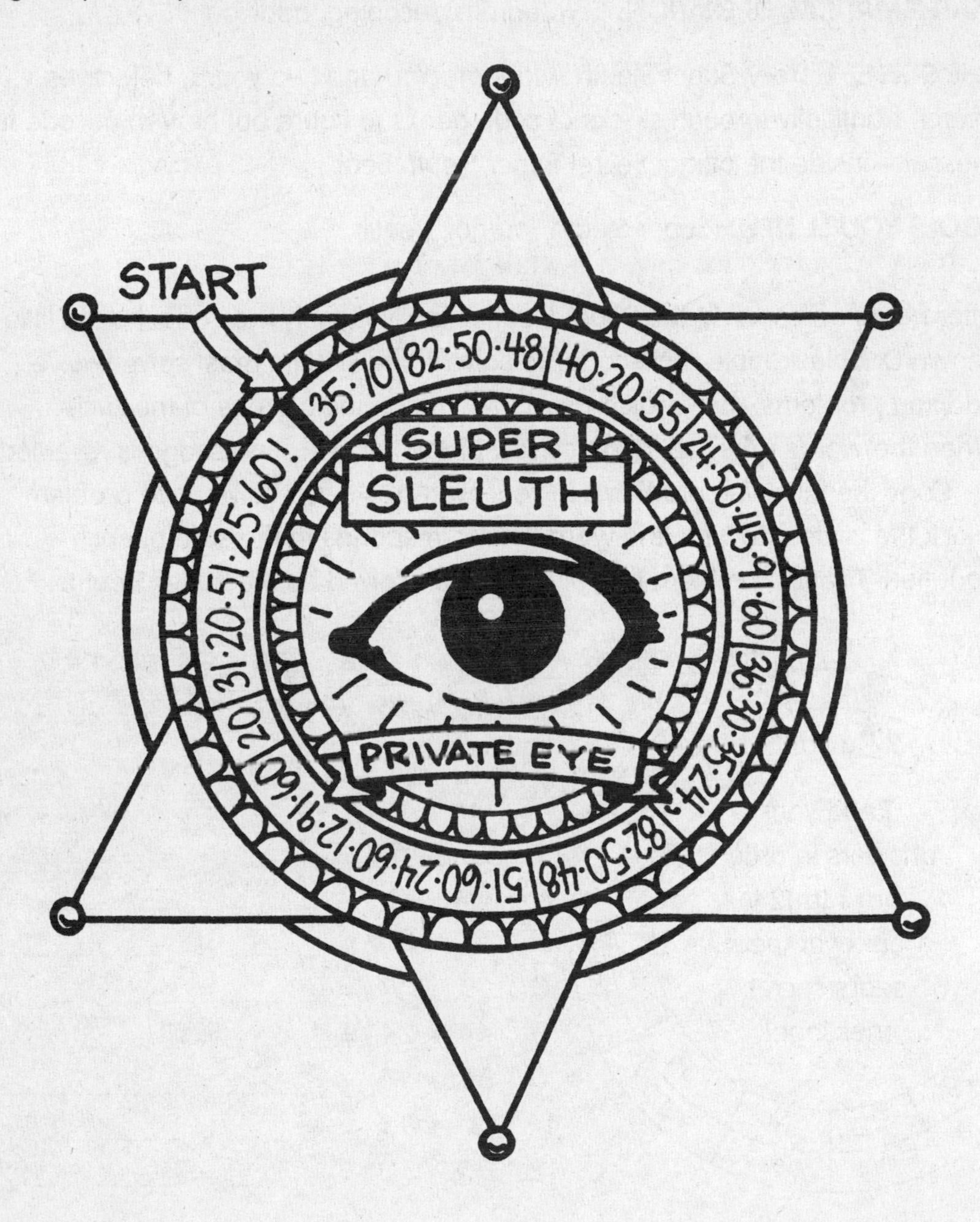

THE CASE OF THE SECRET MESSAGE

MATHEMATICAL OPERATION: Cryptogram decoding; addition

THE SCENE: Library, Super Sleuth Academy. You and two young detectives search frantically through stacks of code books to figure out how to decode the message inside the official Secret Super Sleuth Seal.

TOOLS YOU'LL NEED: Super Sleuth Journal, pencil

After hours of research, the Super Sleuths finally identify the code: It's the little-known Double Trouble circle code. To solve it, the sleuths must solve twelve addition problems, then match them to corresponding areas of the circle. When the words are read in the correct order, the secret message is revealed.

Copy the following problems in your journal. As you solve each problem, check the circle and write the word that corresponds to the sum of each equation. The first problem has already been solved by the Super Sleuths.

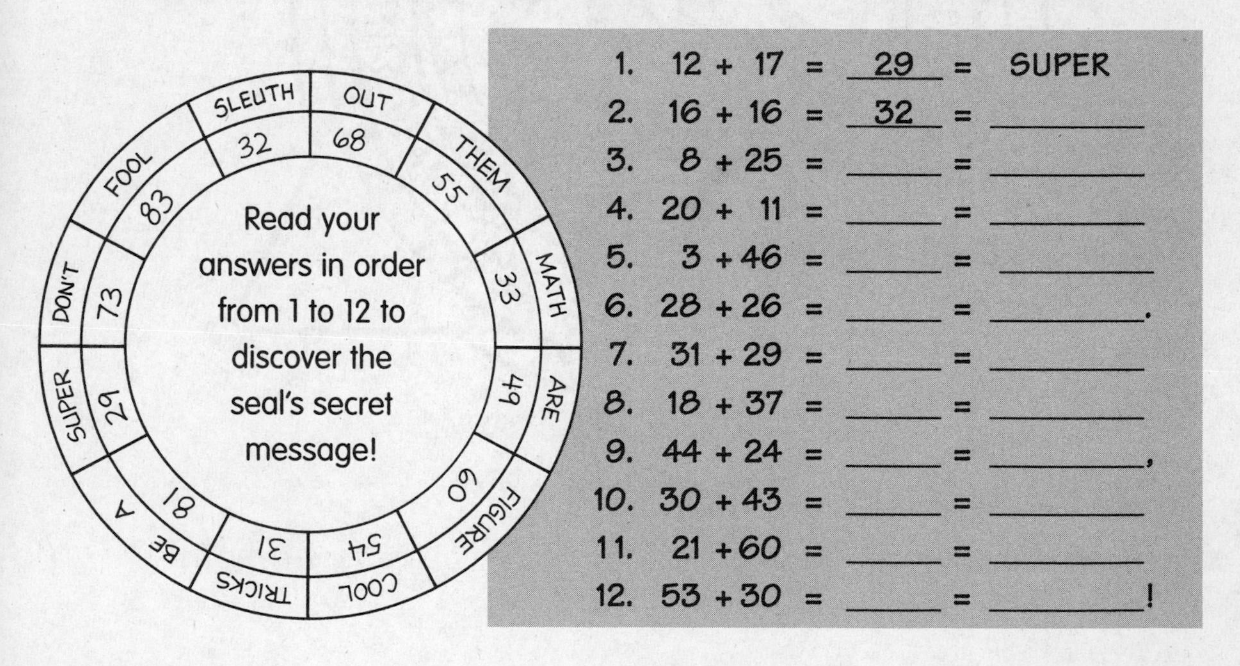

AGENT G.'S MYSTERIOUS CODE

MATHEMATICAL OPERATION: Decoding

THE SCENE: Super Sleuth Academy. A team of cryptographers, including yourself, is puzzling over top-secret information about Agent G.

TOOLS YOU'LL NEED: Super Sleuth Journal, pencil

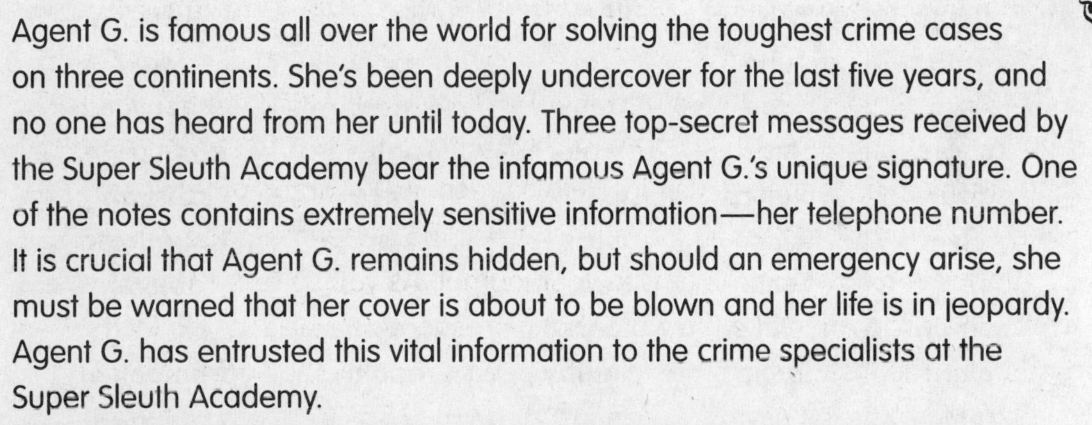

Agent G. is famous all over the world for solving the toughest crime cases on three continents. She's been deeply undercover for the last five years, and no one has heard from her until today. Three top-secret messages received by the Super Sleuth Academy bear the infamous Agent G.'s unique signature. One of the notes contains extremely sensitive information—her telephone number. It is crucial that Agent G. remains hidden, but should an emergency arise, she must be warned that her cover is about to be blown and her life is in jeopardy. Agent G. has entrusted this vital information to the crime specialists at the Super Sleuth Academy.

A team of top-drawer cryptographers assembles to tackle the decoding assignment. Guess what? You made the team! Take a long look at the messages and copy them into your Super Sleuth Journal.

Message 1:

SEPT HUIT NEUF QUATRE TROIS DEUX SIX

Message 2:

SHEE-CHEE HACHI KOO SHEE SAHN NEE ROH-KU

Message 3:

SHEHVAH SHMOENEH TAYSHAH AHRBAH SHAHLOSHE SHATHYEEM HAHMAYSH

As you concentrate on the messages, you realize they're written in different languages. Excitedly, everyone pitches in to create a table of numbers in five languages: English, Spanish, French, Japanese, and Hebrew. Using the table below as a key, you and your fellow sleuths begin to decode Agent G.'s messages by translating them into English.

You have a hunch that the message containing the telephone number will be somehow different from the other two and will stand out as the true message. Only after decoding will you be able to spot the difference.

Key

ENGLISH	SPANISH	FRENCH	JAPANESE	HEBREW
one	uno	un	eechi	achet
two	dos	deux	nee	shathyeem
three	tres	trois	sahn	shahloshe
four	cuatro	quatre	shee	ahrbah
five	cinco	cinq	goh	hahmaysh
six	seis	six	roh-ku	shaysh
seven	siete	sept	shee-chee	shehvah
eight	ocho	huit	hachi	shmoeneh
nine	nueve	neuf	koo	tayshah
ten	diez	dix	joo	ehsehr

SUPER SLEUTH CHALLENGE

Encode a new message using Spanish. Try Agent G.'s telephone number, your own, or one of your best friends'.

REPEAT-A-CRIME CODE

MATHEMATICAL OPERATION: Sequencing

THE SCENE: Super Sleuth Academy, 3 a.m. You and the sleuths are trying to stop the patterned crime-wave made by the Repeat-a-Crime Crooks.

TOOLS YOU'LL NEED: Super Sleuth Journal, pencil

Seven crooks are committing different crimes in the city that follow a pattern. You must find out each crook's next move by figuring out the patterns.

You do some investigative research into past criminal patterns. For instance, one infamous car thief always stole cars according to color and always in the same order: First he'd steal a red car, then a white car, then a blue car, then another red car, and so on. In a similar case, a bank robber only robbed banks on Mondays in January, Tuesdays in February, and Wednesdays in March. He would take the rest of the year off and start all over again the following January.

The sleuths have sketched diagrams that show the patterns of all seven crimes to date. Take a good look at them. Help the sleuths predict the next move and record them in your journal.

1. In a series of jewelry heists, only gems of a certain shape have been stolen. What is the next shape the thieves will try to steal?

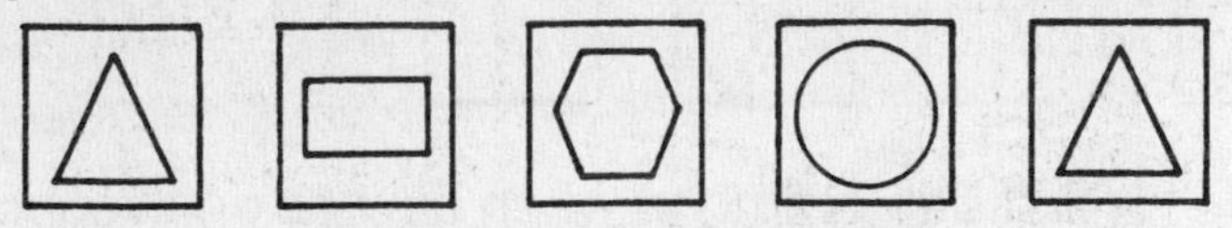

2. A notorious cat burglar hits only certain floors of a luxury office building. Based on the floors he's burglarized in the past, on which floor should the residents lock up their valuables extra tight?

3	2	1	3	2	

3. At an intersection in the middle of town, a pickpocket targets people who are waiting for the "walk" signal to cross the street. Based on the pattern here, on which corner should pedestrians keep both hands on their wallets?

4. The families who live on Poppy Street are hopping mad. Someone has been taking their newspapers before they even have a chance to read them. Based on the addresses that have been hit so far, which family might not get to read the morning funnies?

5	9	14	20	27	

5. Someone's been spray-painting graffiti all over town. Based on the pattern below, what will this vandal's next creation look like?

6. A burglar who specializes in pet stores has stolen several rare lizards from their cages. He takes only a certain number of lizards each time. Based on the pattern below, how many lizards is this loony lawbreaker due to lift?

1	3	2	4	3	

7. Babyface Berletti stole his first baby carriage when he was one year old, a tricycle when he was four, then a mountain bike when he was nine. For him, the urge to steal comes out only every few years. Based on the pattern below, how old will Babyface be when he boosts another set of wheels?

1	4	9	16	25	

HOW DO YOU MEASURE UP?

MATHEMATICAL OPERATION: Ratios

THE SCENE: Fitness room, Super Sleuth Academy. You and other rookie sleuths must figure out your own body measurements so clones can be made.

TOOLS YOU'LL NEED: Super Sleuth Journal, pencil, string or yarn, scissors

As part of the Super Sleuth Academy's physical examinaton, sleuths must give body measurements for the Academy's security records. In this way, a computer-generated clone can be made for each rookie. These clones are used for training exercises that might be too risky for the real rookies.

The first measurements the sleuths must complete are their height and arm span (how much space there is between outstretched arms). Have a friend or a parent cut one piece of string or colored yarn as long as you are tall. Then have him or her cut another piece as long as your arms can reach out. Which is longer? If the two measurements are different lengths, you're a "rectangle." Why? Because if you connect two of your height lengths with two of your arm-span lengths, as in Figure A, you'd create a rectangle shape. If your height length and arm-span length are about the same, you're a "square"—you can connect the four pieces and make a square, as in Figure B below.

A

There are other interesting relationships you can make by measuring your body. For example, there are many ratios between the different lengths of your body parts. A ratio is a comparison between two things. A ratio looks like this: 1:1. In math, the colon (the two periods on top of each other) means that one side is being compared with the other side. If you measure your right foot and get 10 inches,

B

Super Sleuth Challenge

Find the ratio between your heart and your head. How? It's widely accepted that the size of your heart and the size of your fist are a 1:1 ratio. That would mean that if you made a fist, it would be the same size as your heart! So, measure around your fist, then measure around your head. What's the ratio?

and then get 10 inches for your left foot as well, the ratio is 10 inches to 10 inches, written as 10:10. For any comparison, you may reduce it down to the lowest common denominator, as you can do with fractions. In this case, your 10:10 ratio can be reduced down to 1:1. Other ratios don't need to be reduced. For example, if your foot was two times as long as your hand, that ratio would be 2:1 (two hands equal to one foot).

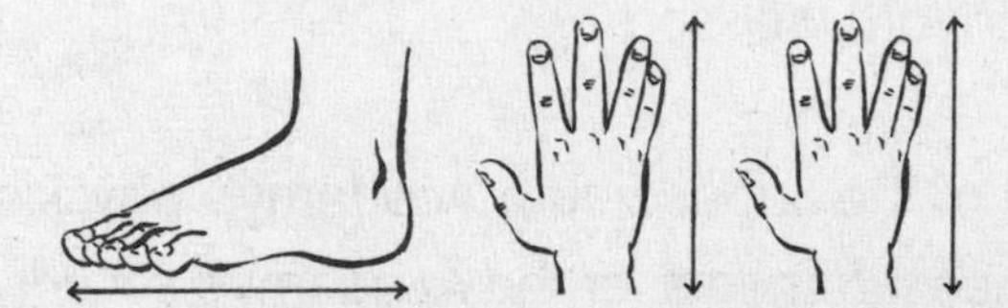

1. Now, use some string to measure the length of your right foot. Compare it with a length of string that goes around your neck. What is the ratio?

Foot: ________
Neck: ________
Ratio: ________

2. Compare your neck measurement with your wrist's. What's the ratio?

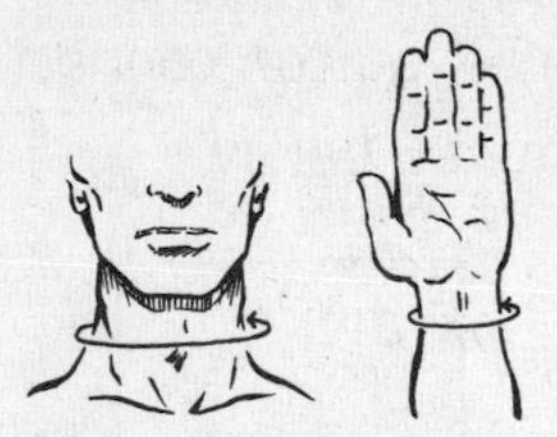

Neck: ________
Wrist: ________
Ratio: ________

3. Compare the length from your elbow to your longest finger with your neck width. What's the ratio?

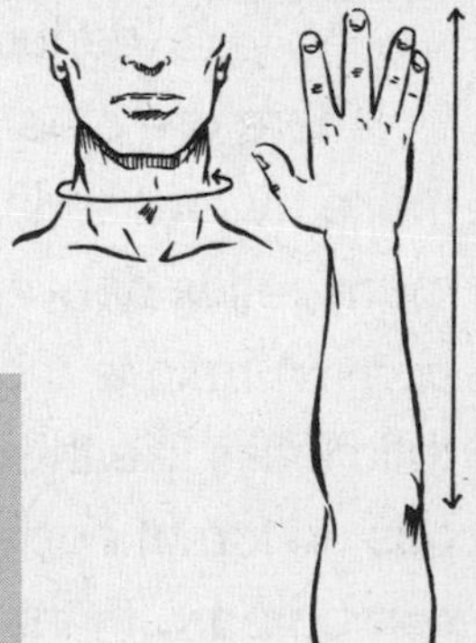

Neck: ________
Elbow to finger: ________
Ratio: ________

FIND THE "FUNNY MONEY"

MATHEMATICAL OPERATION: Pattern recognition

THE SCENE: Next to a counterfeiter's car, parked on a busy street in the city. You and the sleuths need to use the map to locate the spot where illegal printing plates and phony money are hidden.

TOOLS YOU'LL NEED: Super Sleuth Journal, pencil, ruler, colored place markers

In a sack in the trunk of a counterfeiter's car, the sleuths have found what looks like a map of the city laid out block by block. Each square equals one city block. The map covers an area of 100 city blocks. Written on the backside of the map are five important clues. You are sure that this map leads to the hiding place where illegal printing plates and piles of phony money are stashed.

To solve this mystery and find the hidden plates, you'll need your own copy of the counterfeiter's map. Take out your Super Sleuth Journal, a sharp pencil, and a ruler. Copy the map of the city and label each block with the numbers as shown.

1	2	3	4	5	6	7	8	9	10
11	12	13	14	15	16	17	18	19	20
21	22	23	24	25	26	27	28	29	30
31	32	33	34	35	36	37	38	39	40
41	42	43	44	45	46	47	48	49	50
51	52	53	54	55	56	57	58	59	60
61	62	63	64	65	66	67	68	69	70
71	72	73	74	75	76	77	78	79	80
81	82	83	84	85	86	87	88	89	90
91	92	93	94	95	96	97	98	99	100

Now that you have your own copy of the map, pay close attention to the following five clues. They will lead you to the counterfeiter's hiding place. You'll need place markers such as colored candies, small pieces of colored paper, or small pebbles. The first clue is shaded in for you on the chart.

CLUES

1. What does the pattern look like when you find all the numbers with both digits the same? (For example: 55, 22, 33.)
2. What does the pattern look like when you find all the numbers with the digit 5 in them? (For example: 15, 25, 35.)
3. What pattern is made with all the numbers whose digits show a difference of 2? (For example: 42 = 4 - 2 = 2.)
4. Is there a pattern for those numbers with digits that add up to a total of 9? (For example: 81 = 8 + 1 = 9.)
5. Find the pattern with all the numbers that can be evenly divided by 7.

The one *corner block* that was filled in is where you'll find the crooks' hiding place. A few hours later, you and your fellow sleuths raid a certain block number, and bingo! You find hundreds of bags bursting with bogus bucks, along with two snoozing counterfeiters who are using the illegal plates as their pillows! Which city block is it?

HOT OR COLD?

MATHEMATICAL OPERATION: Line graphing, deduction

THE SCENE: Graphing room, Super Sleuth Academy. You and the sleuths plot the location of a notorious spy, Agent Fair N. Hite.

TOOLS YOU'LL NEED: Super Sleuth Journal, pencil, ruler

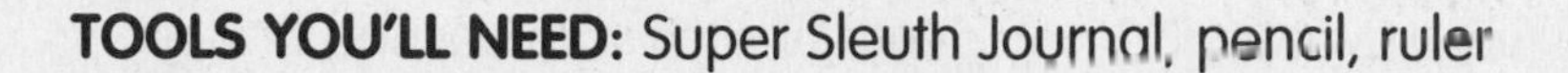

The Super Sleuths are close to capturing elusive spy Agent Fair N. Hite. They've uncovered some data and plotted it on a line graph. From it they should be able to deduce if she is hiding in Hawaii, California, Washington, or Alaska.

In working on this case, the sleuths have discovered several important clues to Agent Fair N. Hite's true identity and her whereabouts. They've learned that she can surf, loves palm trees, and drinks guava punch by the pitcherful.

You can help track Agent Fair N. Hite by taking out your journal, a ruler, and a pencil to copy the graph shown here. Then answer the questions below.

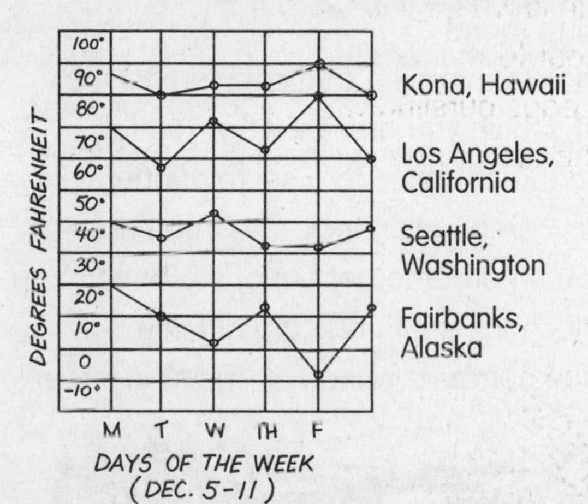

1. Which city is the warmest during the week?
2. Which city is the coldest during the week?
3. Which city has temperatures around the 40s and is cool?
4. Which city has temperatures that are warm, around the 70s?
5. Which two cities might Agent Fair N. Hite choose?
6. Which city will she probably choose and why?

THE RIGHT COMBINATION

MATHEMATICAL OPERATION: Math trick

THE SCENE: Near the entrance of a secret tunnel. You, the sleuths, and Matt "The Math Maniac" must figure out three numbers of a combination lock in order to safely escape the claws of death.

TOOLS YOU'LL NEED: Super Sleuth Journal, pencil, watch with second hand, calculator

Three escaped convicts are holed up in a deserted underground mine where they had stashed a fortune in stolen loot years before. There's no time to waste because the mine is showing signs of caving in, possibly burying the escapees *and* you and your detective buddies chasing them!

Matt stays above ground at the entrance to the mine and can contact the sleuths below via a special sleuth headset, which you are wearing. You tell Matt that you and the group have reached the entrance to a secret tunnel. Unfortunately, it's locked with an enormous combination lock. The three numbers to the combination must be figured out in order to get out of the mine with the stolen money before the whole place turns into one big dirt pile.

One sleuth finds some clues in a series of numbers written on a wall in three columns. Matt believes that the totals from each of the three columns will give the sleuths the secret three-number combination. There's no time to lose! Quickly, you read off the numbers. Matt smiles as he recognizes the numbers from a past mystery. He uses a special math trick that works especially for these three math problems.

Here's how to do it. First write down all three columns of numbers in your journal:

COL. 1	COL. 2	COL. 3
6497	5672	5624
5852	4727	6979
9725	3848	3486
+ 7649	+ 9599	+ 7395

Pull out a watch with a second hand and add the first addition problem as fast as you can. You're allowed to use a calculator if you want. How long did it take you? Here's how Matt added the first column of numbers.

Step 1: Take the third number down from the top:	9725
Step 2: Subtract 2:	9725 - 2 = 9723
Step 3: Write a 2 in front of the new number:	29723

29723 is the answer! Check it against yours.

Try this trick on the remaining two columns and write the three sums in your journal.

Once you learn Matt's secret, you quickly relay the correct sums to the other sleuths. A young rookie grabs the stolen loot, and you lead everyone to safer ground. Just in time! Suddenly, the mine caves in and sends a huge cloud of dust and dirt high into the air.

Super Sleuth Challenge

Because this trick only works for these particular numbers, you can't use this as an addition shortcut. However, why not turn it into a math trick with which to amaze your friends? Simply put the three addition problems on separate sheets of paper. Have a friend pick one, but not show you. Then choose another friend and give him or her a calculator. On the count of three, your friend should hold up the problem for you and your other sleuthy buddies to solve. Who is the quicker math genius? You'll beat the calculator every time!

THE STEPPING-STONE NUMBER CHASE

MATHEMATICAL OPERATION:
Math trivia

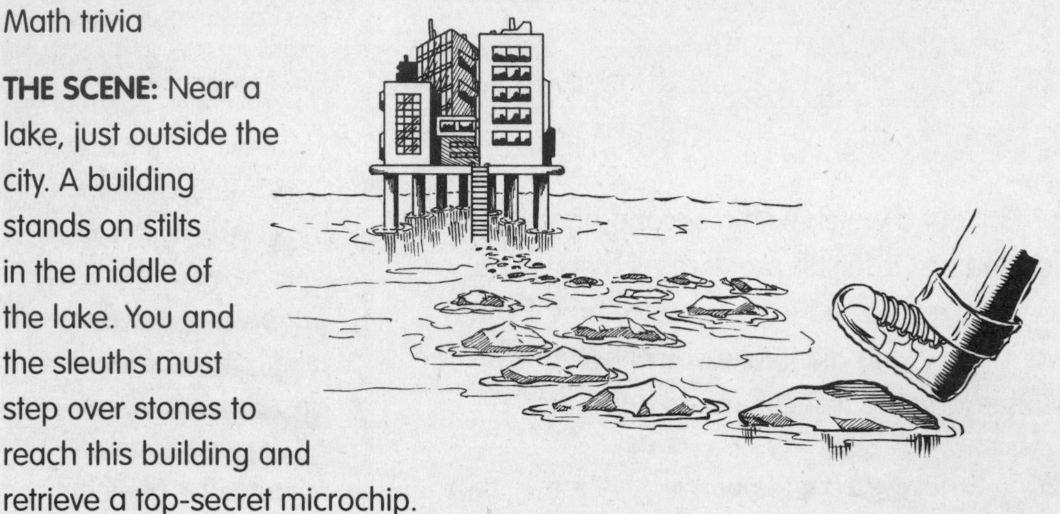

THE SCENE: Near a lake, just outside the city. A building stands on stilts in the middle of the lake. You and the sleuths must step over stones to reach this building and retrieve a top-secret microchip.

TOOLS YOU'LL NEED: Super Sleuth Journal, pencil

The Super Sleuths are hot on the trail of a sly, squirmy suspect who has stolen a top-secret microchip capable of revolutionizing computers everywhere. He's swiped the chip from a high-tech company that was originally built on stilts in the middle of a lake for added security. The only way to cross the lake is to follow a trail of numbered stepping-stones that lead safely to land. There's just one little problem: some of the stones are electrically charged, so that those who step on them get a fatal case of hot foot!

Luckily, one of your partners has located a security file showing which stones are charged and which are not. The security guard must have been a big fan of math trivia, however, because each stepping-stone has a number that corresponds to a math trivia question. If the answer to a question is an even number, look out! The stone is charged! If the answer to a question is an odd number, you may step on the stone.

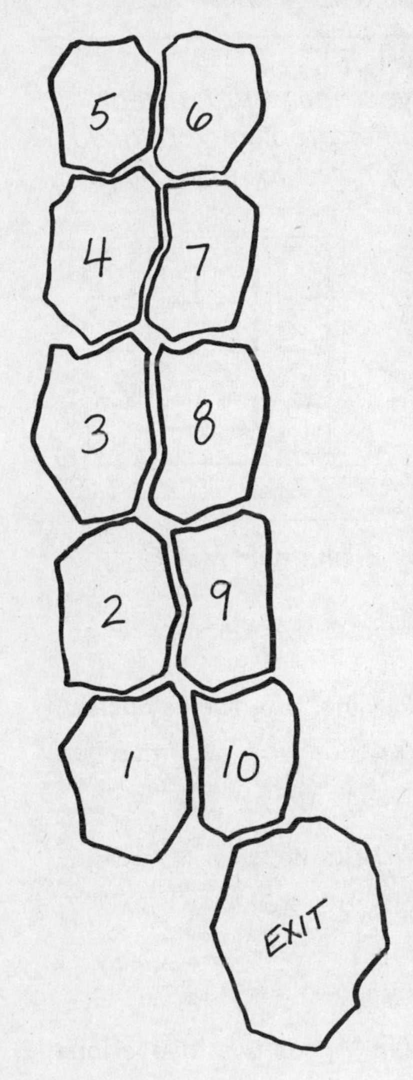

Take out your journal and write each answer down. Decide which stones you would step on (odd number answer) and which you'd hop over (even number answer.)

Answer the following questions and decide whether to step or not.

1. One decameter is equal to how many meters?
2. How many months are there in a year?
3. What number lies exactly between 20 and 30?
4. What's the difference between 53 and 42?
5. How many sets of 10 are in 100?
6. How many dimes are there in $1.00?
7. What is one less than 4 multiplied by itself?
8. What's the lowest multiple of 3 that is evenly divisible by 7?
9. How many doughnuts are in a box of two dozen?
10. How many minutes are there in one day?

Check your answers in the back to see if you made it across and collared the chump who ripped off the chip!

Math Sleuth Rookie QUIZ

Now it's time to test what you've learned. Open your journal to a clean page. Each quiz question corresponds with an earlier problem you faced in this chapter. If you get stumped, turn to the mystery noted for extra help.

1. In "Get the Facts," you helped the Super Sleuths sharpen their skills in word problems. Now try this one:

   Step 1: Think of your home address–just the number, not the street.

   Step 2: Add 3, and multiply the sum by 2.

   Step 3: Add 4 to the product. Divide the sum by 2, and subtract 5.

   What do you get?

2. In "Badge Me," you helped the sleuths tackle decoding. Use the key below to fill in the letter on the dotted line that corresponds with each number. What's the message?

   Key: A =32 R =12 G =41 E =23 T =5 W =56 K =47 O =8

   41 - 12 - 23 - 32 - 5 56 - 8 - 12 - 47

   ___ ___ ___ ___ ___ ___ ___ ___ ___

3. Do you remember how to figure out the solution in "Agent G.'s Mysterious Code"? Use the code on page 12 to decipher the telephone number below.

   uno - deux - goh - shahloshe - cuatro - nine - achet

   ___ ___ ___ ___ ___ ___ ___

4. How good were you in calculating the sequencing patterns in "Repeat-a-Crime Code"? Here's another pattern to complete:

   0 - 1 - 3 - 6 - 10 - 15 - ?

5. Did you find the answer to "The Case of the Secret Message"? Use the following key to decode the message below.

Key:

A=1	H=3	O=6	T=8
C=13	I=4	P=7	U=9
E=2	L=16	R=12	W=10
F=15	N=5	S=17	Y=11

4 10-1-5-8 11-6-9 8-6 12-2-1-13-3 8-3-2 8-6-7 6-15 11-6-9-12 13-16-1-17-17!

6. In "How Do You Measure Up?" you helped the sleuths sharpen their skills in measuring. Without a ruler, figure out which is longer, your hand or foot? How about your foot or around your neck? Your nose or little finger? Your arm or leg?

Are there any 1:1 ratios? How about 2:1 ratios?

1	2	3	4	5
6	7	8	9	10
11	12	13	14	15
16	17	18	19	20
21	22	23	24	25

7. "Find the 'Funny Money'" showed you how to recognize different patterns. Do the following steps, then answer the question below. First shade all the odd numbers, then shade all numbers divisible by 3.

How many numbers remain?

8. "The Stepping-Stone Number Chase" tested your M.T.Q. (math trivia quotient). Here are a few more math questions to solve. What's the common number in all the answers below?

a) Number of days in two weeks: ____

b) Number of eggs in a dozen: _____

c) The next century: _____

d) Difference of 21 and 10: ____

Check your answers against those listed on page 62. How well did you score?

2: Math Sleuth Cadet

THE CUTUP CAPER

MATHEMATICAL OPERATION: Geometry, three-dimensional visualization

THE SCENE: Stan's Supermarket. You and your sleuth buddies must investigate the damages from a daring midnight robbery.

TOOLS YOU'LL NEED: Paring knife, a rib of celery, a banana, and other various fruits and vegetables such as an apple, an orange, or a carrot

Stan, the owner of the vandalized supermarket, needs the investigation done quickly so he can report his losses to the insurance company. Pieces of fruits and vegetables are splattered on the walls, shelves, and bins of poor Stan's once-tidy store.

One look at the expertly sliced produce leads the sleuths to suspect a thief named Curt "The Cutup" Carnahan. He's vandalized the fruits and vegetables sections of many markets by cutting up apples, bananas, celery, carrots, and potatoes in different angles. To give the insurance company an estimate on how much harm he's done, the sleuths will have to inventory the damage. You can help them find what's been destroyed by cutting up a rib of celery and studying the different shapes you can make just by adjusting the angle of the cut. Be sure to have an adult help you do this. Look at the illustration to see which angles should be cut.

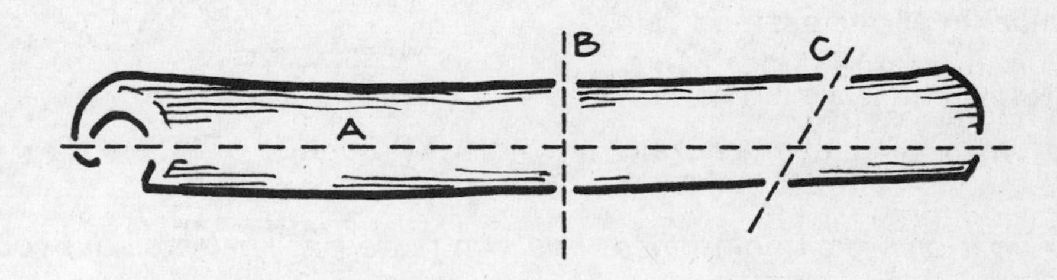

After you've made the three different angles, see if you can match up the slices with their shapes:

If you matched up A with Shape 1, B with Shape 2, and C with Shape 3, you're on your way to becoming quite a "cutup" yourself. Now, with the help of an adult, try cutting up this banana. Look at the illustration to see how to slice the banana.

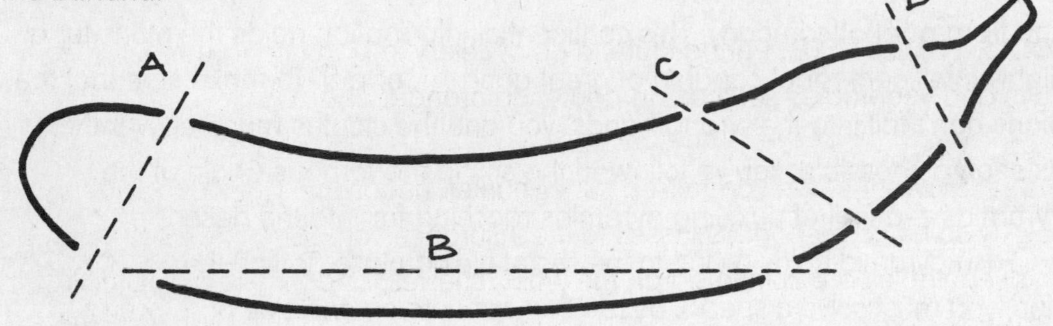

Can you match up these slices with their shapes?

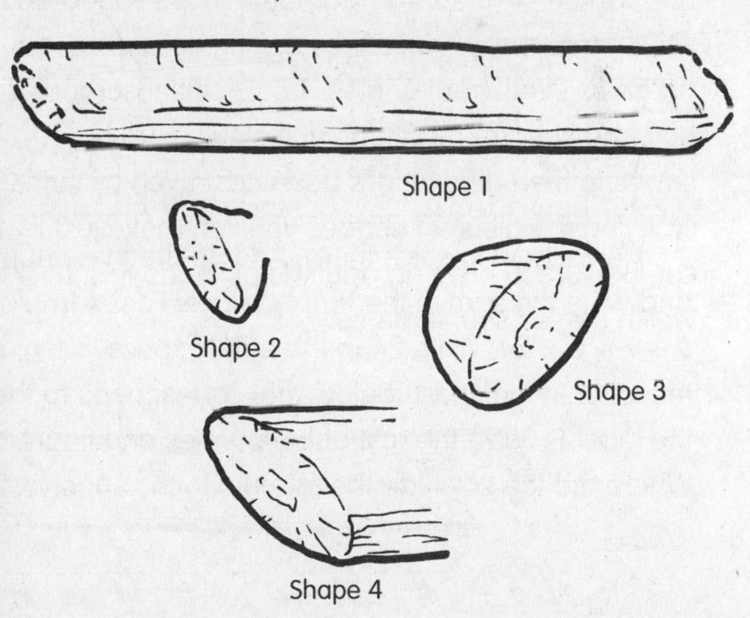

SUPER SLEUTH CHALLENGE

What other fruits and vegetables can you slice to make different shapes? (When using a knife, always have an adult help you.)

THE PYRAMID'S CHAIN

MATHEMATICAL OPERATION: Addition, cryptography

THE SCENE: The Egyptian desert. You and the sleuths must find a spy who is capable of creating an evil robot.

TOOLS YOU'LL NEED: Super Sleuth Journal, pencil

The Super Sleuths are in Egypt to find a spy who has stolen top-secret information from a robot company. This confidential information holds the plans for a highly intelligent robot capable of great good . . . or evil. To make sure that the plans don't fall into the wrong hands, you and the sleuths must recover them as soon as possible. You've followed the spy to the famous Chain of the Pyramids—a trail of puzzling pyramids reaching far into the desert.

Each pyramid holds a clue to the secret hiding place. To find the clue, you must solve a special puzzle that appears on one side of each pyramid. Copy the pyramid puzzles below into your journal.

7
9 8
P

10
5 7
E

13
4 1
H

8
10 2
M

14
15 5
L

7
6 3
U

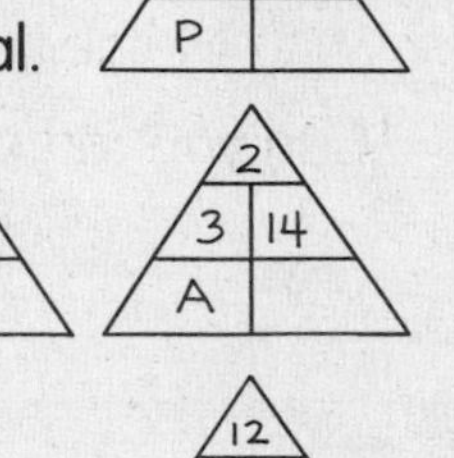

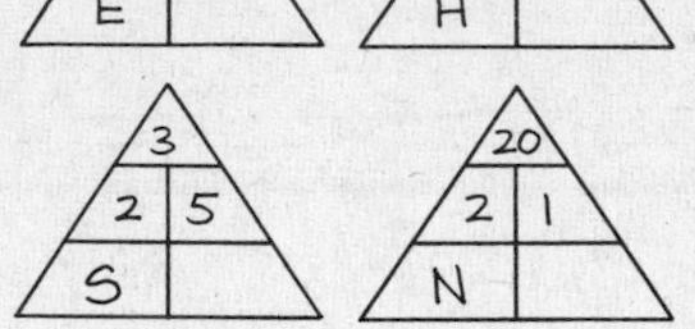

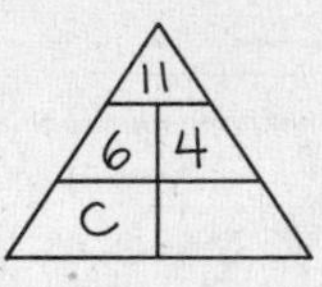

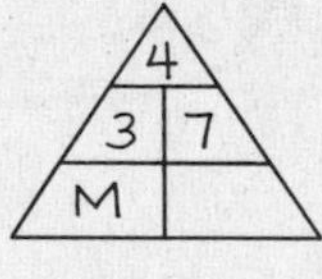

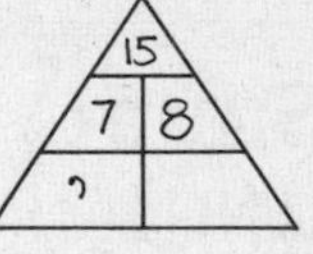

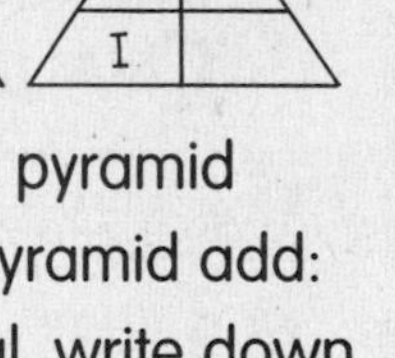

Here's how to solve them: Add up the three numbers in each pyramid and write the sum in the empty space. For example, in the first pyramid add: 9 + 7 + 8 = 24. Write 24 in the empty space. Then, in your journal, write down the letter in the space below that corresponds to the answer. In this case, it's the letter P. Solve the rest of the puzzles and figure out the coded message. Where did the spy hide the formula to the super-colossal robot?

__ __   __   __ __ __ __ __ __ __   __ __ __ __

27 23   19   21 19 20 22 34 30 10   18 16 14 24

THE CLEAN GETAWAY

MATHEMATICAL OPERATION: Topology

THE SCENE: Inside a factory warehouse. You and two other sleuths must find a way through a maze of boxes to find four dangerous criminals.

TOOLS YOU'LL NEED: Super Sleuth Journal, pencil

Sleuths know how important it is to find a solution in the fewest possible steps. Retracing their steps can mean going on . . . and on . . . and on, while the crook gets away! Two Super Sleuths stand outside a factory warehouse where four dangerous criminals may or may not be hiding. Inside the warehouse are thousands of crates and boxes stacked to the ceiling. These crates and boxes make it almost impossible to get from one end of the warehouse to the other. In fact, the inside of the warehouse looks like a maze.

With the help of the factory's owner, the sleuths draw a map of the warehouse floor and discover that the four different crooks spread out and took four different paths in the shapes shown here.

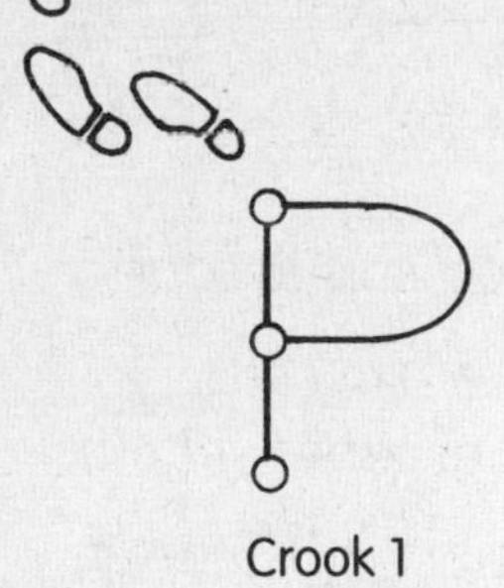

Crook 1

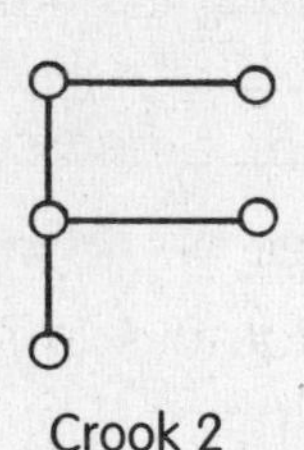

Crook 2

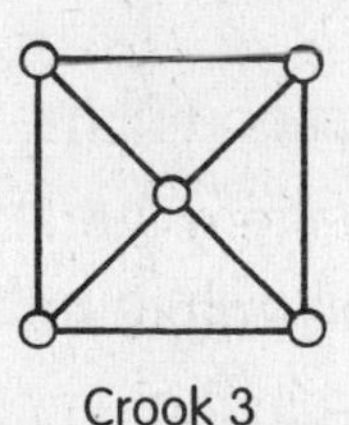

Crook 3

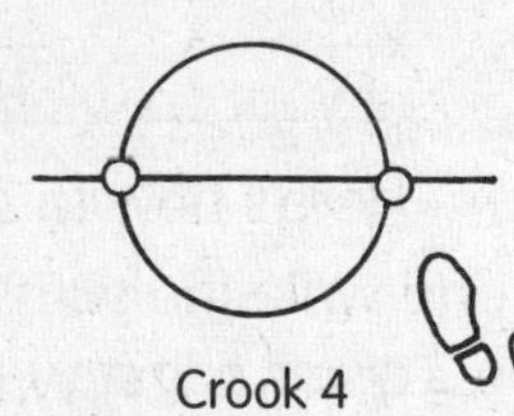

Crook 4

After studying the mazes, the sleuths instantly realize that this case involves a branch of mathematics called *topology*—the study of surfaces and shapes. They know they must identify three special places on each shape: the vertex, the enclosed region, and the arc.

The facts:

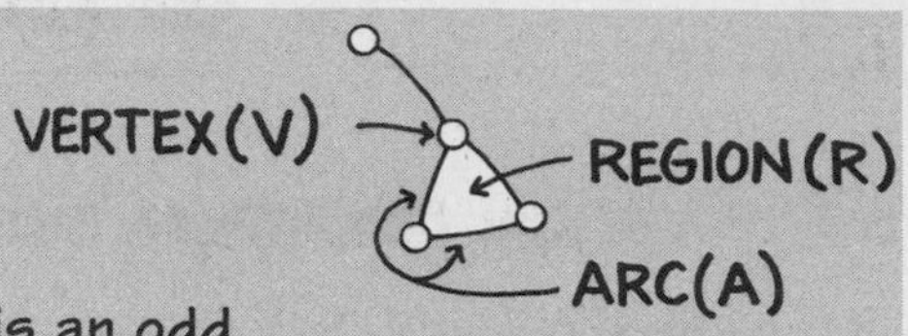

- The *vertex* is the point where two or more lines meet or intersect. The plural of vertex is *vertices*.

VERTEX(V)
REGION(R)
ARC(A)

- If the number of lines meeting together is an odd number, like 3 or 5, the shape has an odd number of vertices.
- If the number of lines meeting together is an even number, like 2 or 4, the shape has an even number of vertices.
- The enclosed region is the space completely surrounded by lines.
- An *arc* is the area between two vertices.

Look at the shape mazes again. Copy them in your journal and find the number of enclosed regions, arcs, and vertices for each one. By using this information in a formula (a step-by-step way to solve a mathematical problem), you can find out if there is an escape route out of any maze! Here's the formula for figuring the four escape routes: $(V + R) - 1$

The formula tells you to:

1. Add the number of vertices (V) together.
2. Add the number of enclosed regions (R) together.
3. Add the number of vertices and regions together $(V + R)$.
4. Take your answer and subtract 1 to get the solution $(V + R) - 1$.

If your solution is an odd number, like 3, 5, 7, and so on, you can escape without retracing a line.

If your solution is an even number, like 2, 4, or 6, you can't escape without retracing a line.

The formula $(V + R) - 1$ solves the puzzle for you so you don't even have to put on your shoes. Which of the four mazes offer no escape without retracing steps? In your journal, write down your answers to the escape routes of Crook 1, Crook 2, Crook 3, and Crook 4.

SHAPE IT UP

MATHEMATICAL OPERATION: Distinguishing and sorting shapes

THE SCENE: Super Sleuth Academy's dining room. Detective Smith helps you and the other sleuths find the location of precious stones that have been hidden.

TOOLS YOU'LL NEED: Super Sleuth Journal, pencil, ruler

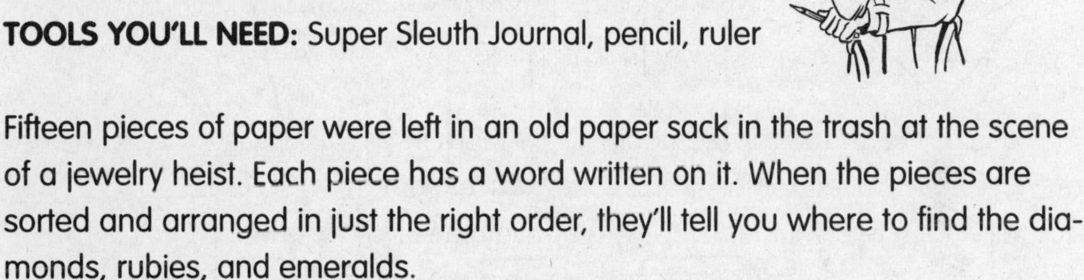

Fifteen pieces of paper were left in an old paper sack in the trash at the scene of a jewelry heist. Each piece has a word written on it. When the pieces are sorted and arranged in just the right order, they'll tell you where to find the diamonds, rubies, and emeralds.

The sleuths have brought the odd scraps of paper to the Academy's dining room, where they can be spread out on a table. One of their friends and teachers, Detective Smith, is helping them solve the mystery by using his Double Circle trick. You can help them find the solution by getting out your Super Sleuth Journal and a sharp pencil to follow along as the detective sorts the weird shapes into two groups. First he draws two large circles on a piece of white paper. Then he takes all the pieces to be sorted and puts them outside of the two circles. Sound easy? Let's see if you can do it.

The shapes found at the scene of the crime are all quadrilaterals (four-sided figures). But they are slightly different: some figures are parallelograms, and some are not parallelograms. How can you tell? Parallelograms are quadrilaterals whose opposite sides are parallel and equal, like the three examples below:

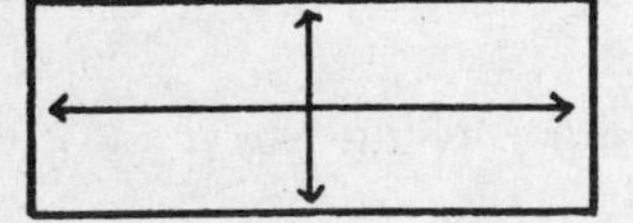

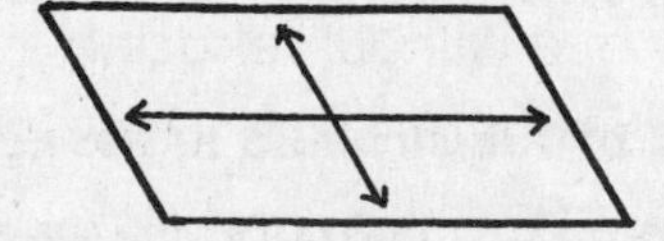

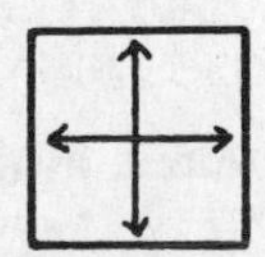

Help the sleuths group the figures according to which are parallelograms and which are not. In your journal, draw the parallelograms in Circle 1 and the non-parallelograms in Circle 2.

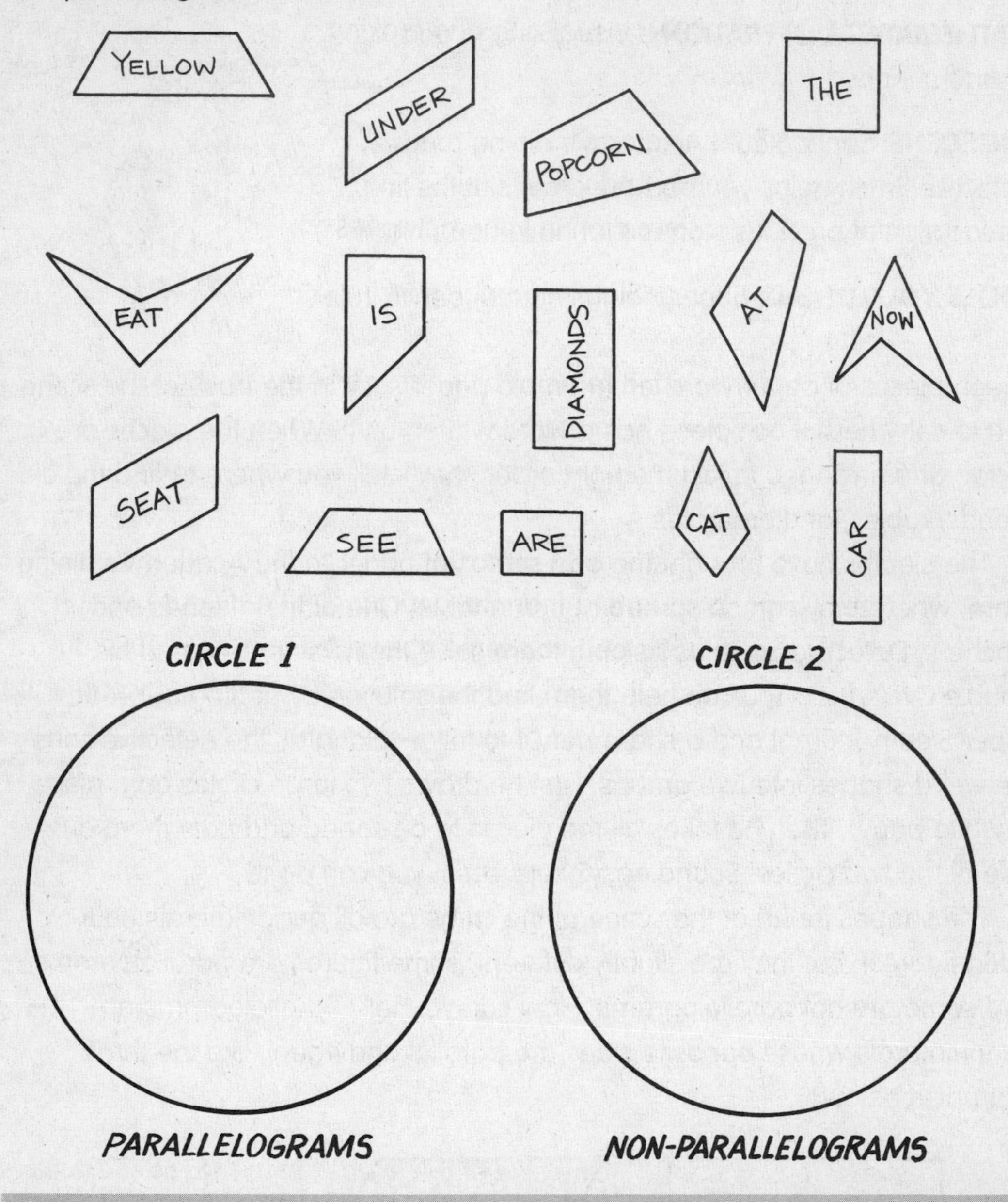

The words inside one of the circles spell a message telling where to find the lost jewels. Is the message in Circle 1 or Circle 2?

MEMORY MADNESS

MATHEMATICAL OPERATION: Visualizing and making shapes, estimating sizes

THE SCENE: Memory training class, Super Sleuth Academy. You and the sleuths are practicing to retain information in your memories.

TOOLS YOU'LL NEED: Super Sleuth Journal, pencil

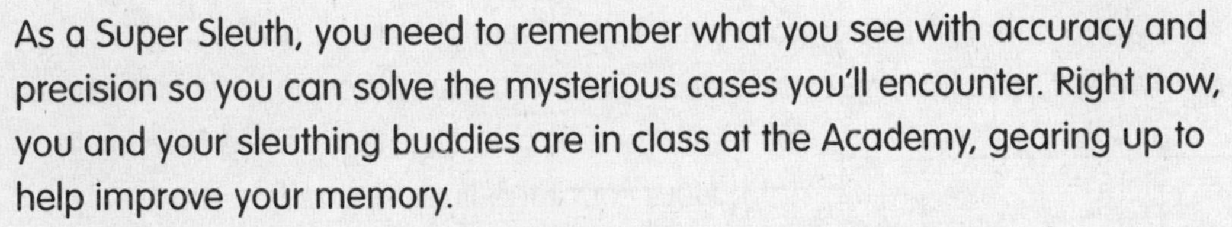

As a Super Sleuth, you need to remember what you see with accuracy and precision so you can solve the mysterious cases you'll encounter. Right now, you and your sleuthing buddies are in class at the Academy, gearing up to help improve your memory.

Close your eyes. Can you picture a hexagon? It's the shape of a stop sign. Now visualize a circle, a square, a triangle, a rectangle. Pretty easy assignment, huh? Go a step further. Can you remember sizes as easily? Take out your Super Sleuth Journal and a pencil, then with your eyes closed, picture the five items below one at a time. Next open your eyes and try to draw the *actual size* of these everyday shapes.

1. Draw a square the size of the push button on a telephone.
2. Draw a circle the size of a new roll of toilet paper.
3. Draw a rectangle the size of a dollar bill.
4. Draw a triangle the size of a large waffle ice-cream cone.
5. Draw a circle the size of the top of a soda-pop can.

Match up your "guesstimate" of the size of the shapes with the real objects. For example, match the picture you drew of a push button to the actual push button on a telephone. The more you practice, the better you'll get.

THE VANISHING CROOK

MATHEMATICAL OPERATION: Line paradox

THE SCENE: Lineup room, police headquarters. You must help the Super Sleuths find a criminal who has vanished right before your eyes.

TOOLS YOU'LL NEED: Super Sleuth Journal, pencil, ruler, scissors

A crook in a lineup at police headquarters seems to disappear and reappear. Can you help the Super Sleuths solve this baffling puzzle?

Take out a sheet of paper from your Super Sleuth Journal, and using a ruler, draw ten lines like the ones below. Cut the paper diagonally as shown, then shift the left side slightly upward toward the right. Now count the lines—one line magically disappears! Now shift the halves back to their original positions and watch the line reappear.

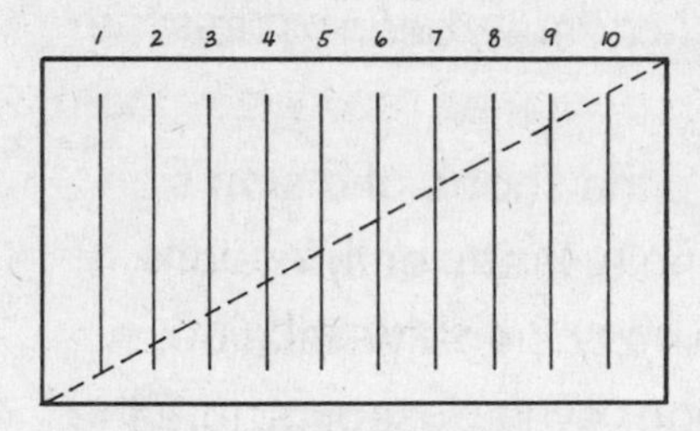

This is called a line paradox. A paradox is something that seems to happen, but really doesn't—just like the lines when you shifted them from left to right and one seemed to disappear, then magically reappear. Now copy the faces below on another piece of paper in your journal. Be careful to get the mouth, nose, eyes, and hats in the right places. Then take out your scissors and cut on the dotted line. When you slide the lower piece to the left, the line paradox happens again. Count the faces. What happened?

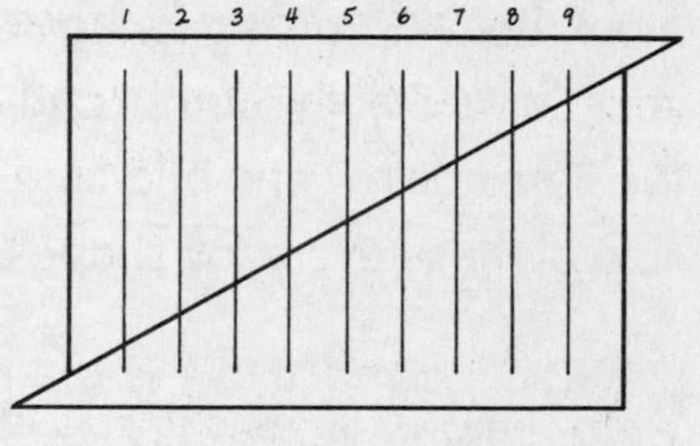

TANGRAM TWINS

MATHEMATICAL OPERATION: Tangrams

THE SCENE: Police headquarters. You and the sleuths must find where some stolen money is hidden.

TOOLS YOU'LL NEED: Pencil, ruler, one piece of construction paper

While searching a bank robbery suspect, the Super Sleuths find several crumpled pieces of paper with numbers written on them. Since the suspect had only fifty cents in his pocket, you figure these paper scraps might tell where the stolen money is hidden. The only way to be sure is to decipher the mixed-up message.

When one of the sleuths uncrumples the pieces of paper, he recognizes them as parts of a tangram. A tangram is a puzzle made up of several different shapes, like triangles, squares, and parallelograms. To reconstruct this puzzle and figure out its message, use a ruler and pencil to copy the shapes below onto a piece of construction paper. Be sure to write the numbers on each shape.

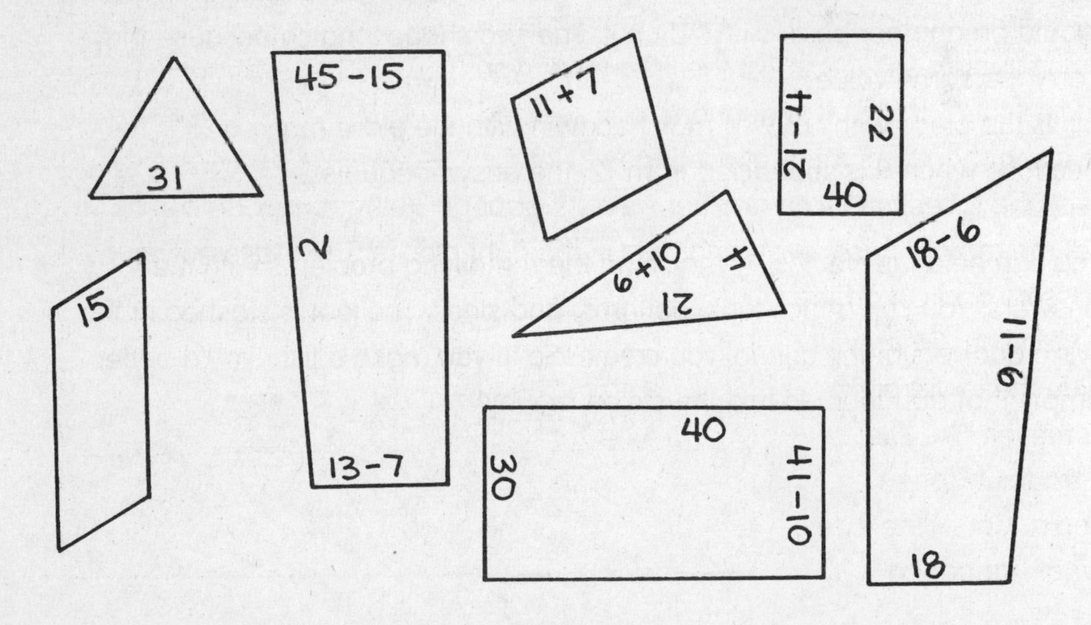

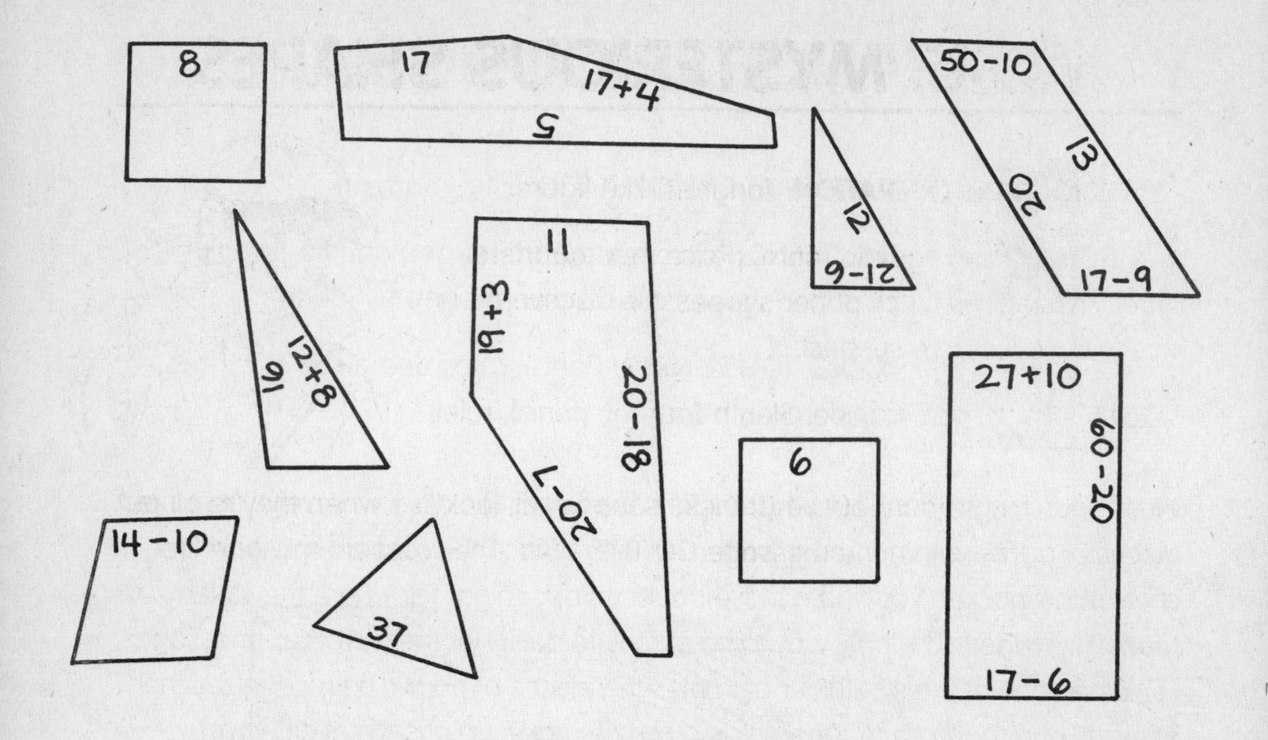

After you copy the numbered shapes, cut them out. Match up the shapes by recombining the pieces together so a problem is matched with its answer.

Here's how to do it: If a piece has "11-7" on it, a possible match, or "twin," would be another piece with "4" on it. The two shapes match because they carry the same value.

Is the piece with "12-9" a matched twin with the piece marked "3"? Yes, because when 9 is subtracted from 12, the answer equals 3.

You can help the sleuths by matching the remaining problems with their answers. You can match tops, bottoms, and sides. The loot is stashed in the living quarters of the animal you create. So, if you make a fish, you'd better empty that aquarium to find the stolen money!

MORE MYSTERIOUS SHAPES

MATHEMATICAL OPERATION: Tangram-like figures

THE SCENE: Interrogation room, police headquarters. Three mysterious black paper shapes are discovered in the coat pocket of a suspect.

TOOLS YOU'LL NEED: Super Sleuth Journal, pencil, ruler

Help the sleuths figure out what these shapes will look like when they're all put together so they form the message GO THIS WAY. The word *go* must be first, *this* must be next, and *way* must be last. Draw the images in your Super Sleuth Journal.

GO

THIS

WAY

When you put the three clues together in the correct order, which object below does it resemble?

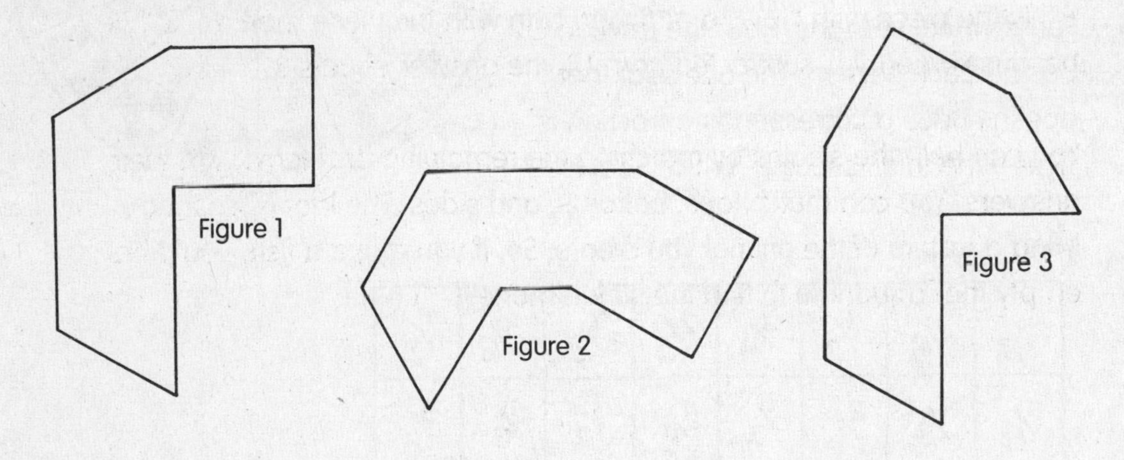

THE FRACTURED FRACTION FERRIS WHEEL

MATHEMATICAL OPERATION: Fractions

THE SCENE: Country fair. The Super Sleuths have chased an art thief to a country fair.

TOOLS YOU'LL NEED: Super Sleuth Journal, pencil

In this case, you must help the sleuths find the whereabouts of an art thief and her priceless painting. She's hiding somewhere on the Ferris wheel. Copy the fraction wheel in your Super Sleuth Journal, then look at the fraction clues in the boxes below. Try to find fractions in the key that match each of the drawings in the Ferris wheel sections. Fill in the fraction beside each section. The one seat that doesn't have a corresponding fraction is where the thief is hiding. Which seat is it?

Key

1	2	3	4	5	6	7	8
$1/3$	$1/4$	$1/2$	$2/4$	$2/5$	$2/3$	$3/6$	$1/2$
9	**10**	**11**	**12**	**13**	**14**	**15**	**16**
$1/4$	$1/3$	$2/6$	$1/2$	$4/8$	$1/2$	$1/6$	$3/8$

WAISTLINE BANDITS

MATHEMATICAL OPERATION: Topology

THE SCENE: Department store, crime scene. You've found three belts left in various parts of the store where crimes were committed. But who do they belong to? The Chief has rounded up five groups of suspects, and you must identify the true criminals and the crimes they committed. Otherwise, everyone gets off free!

TOOLS YOU'LL NEED: Super Sleuth Journal or extra notebook paper, pencil or marker, scissors, ruler, tape

The lineup includes:

1. Jim and John, the look-alike twins.
2. Sue and Sassy, the inseparable twin sisters.
3. Fat Fred, the overweight crook.
4. Slim Sam, the toothpick crook.
5. Bill, Bud, and Bob, the tricky triplets.

To help with the investigation, you'll need to examine the evidence closely. Here's how to make your own paper versions of what the three belts looked like. Take out the tools listed above and cut six to eight strips of notebook paper about 1 inch wide.

Belt 1 was found in the toys and games section. Here's how to make it:

1. Tape two strips together at the ends. It should look like this:

2. Starting near the tape, draw a line down the middle with a pencil or marker and cut the belt along the middle line of the paper. What happens? Whose belt (or belts) do you suppose it might be?

Super Sleuth Challenge

Belt 2 is called a Möbius strip, named after the famous nineteenth-century topologist August Möbius. What other things can you make with a Möbius strip?

Belt 2 was found in the jewelry department. To make it, follow the directions below.

1. Tape together two more strips of paper.
2. Before you tape the ends together to make it a complete belt, give it a half twist—twist only one side—then tape the ends.

3. Now draw a line down the middle and cut along the middle line again. What happens now? Which suspect might have worn this belt (or belts)?

Belt 3 was found in footwear. Here's how to make belt 3.

1. Tape two strips together. Before taping the ends, give one end of the strip a full twist around and tape the ends tightly.

+ (TAPE) + FULL-TWIST

2. Draw the line along the middle and cut along the middle line again. What happens now? Who could have worn a belt (or belts) like this?

Take the belts and match them with their owner or owners. Who are the guilty parties?

MATH SLEUTH CADET QUIZ

Now it's time to test what you've learned in chapter 2. Open your journal to a clean page. Each quiz question corresponds with a problem from this chapter. If you get stumped, turn to the mystery noted for extra help.

1. Using your decoding skills from "The Pyramid's Chain," use addition to decode where you can find the hidden jewel.

   ___ ___ ___   ___ ___ ___
   10 24 19   30 22 21

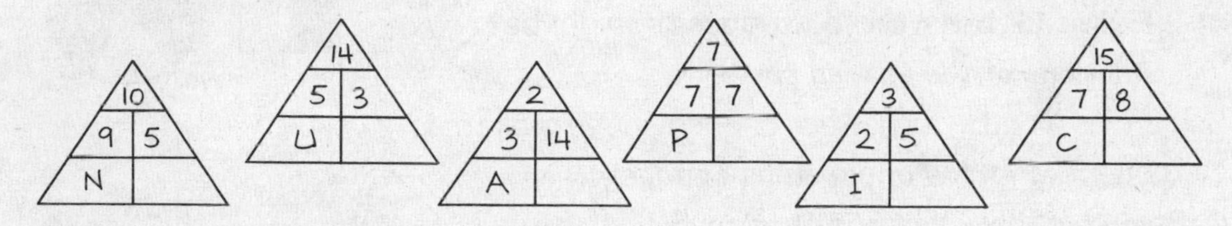

2. In "Shape It Up" you separated shapes into two groups, parallelograms and non-parallelograms. To uncover this latest message, arrange parallelograms in Circle 1 and non-parallelograms in Circle 2. Which message makes the most sense?

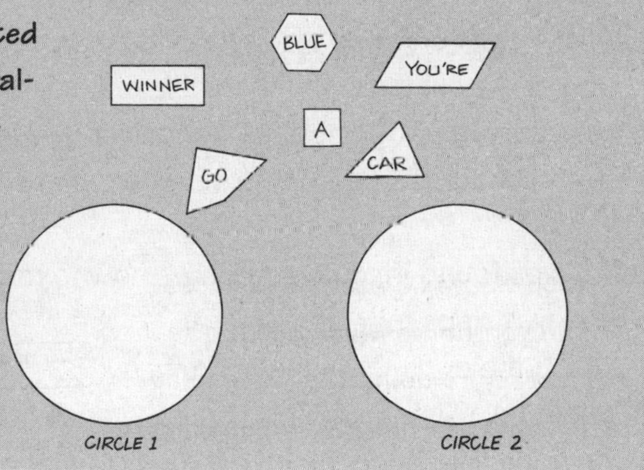

3. How good were you at estimating sizes in "Memory Madness"? Draw the objects listed below.

   1. a stapler
   2. a dime

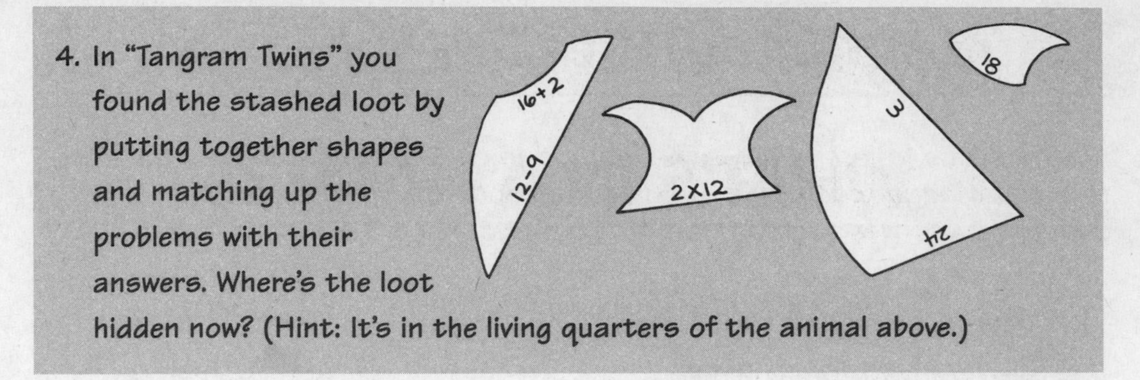

4. In "Tangram Twins" you found the stashed loot by putting together shapes and matching up the problems with their answers. Where's the loot hidden now? (Hint: It's in the living quarters of the animal above.)

5. Do you remember how to recognize the fraction shapes from "The Fractured Fraction Ferris Wheel"? What are the fractions of the shaded parts?

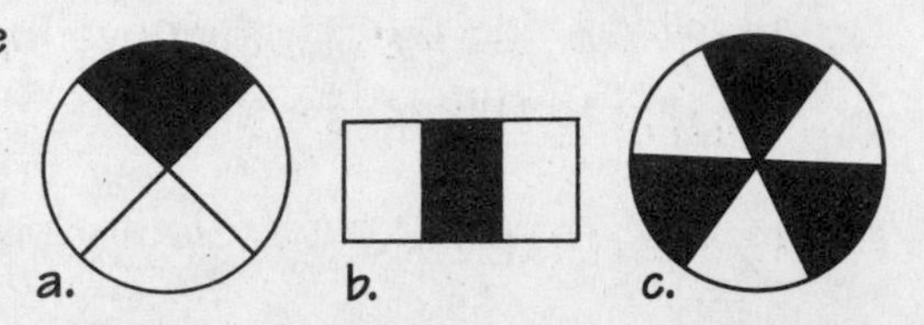

6. With two strips of paper taped together and cut down the middle, what figure shown here do you get? (For help, refer to "Waistline Bandits.")

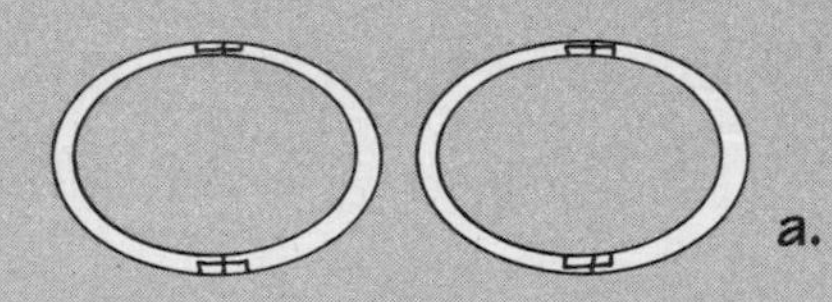

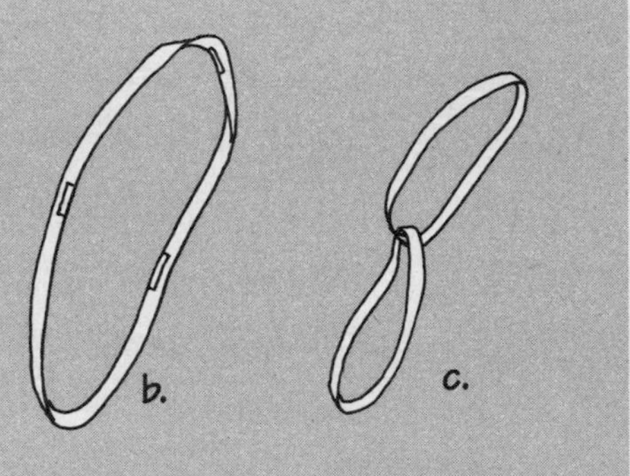

7. With "More Mysterious Shapes," can you figure which one is the final figure using the three shapes below, and what its message is?

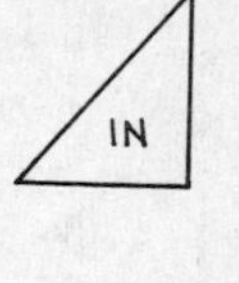

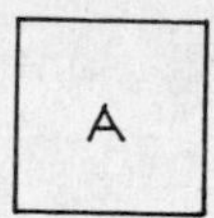

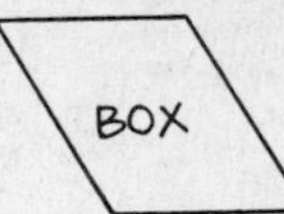

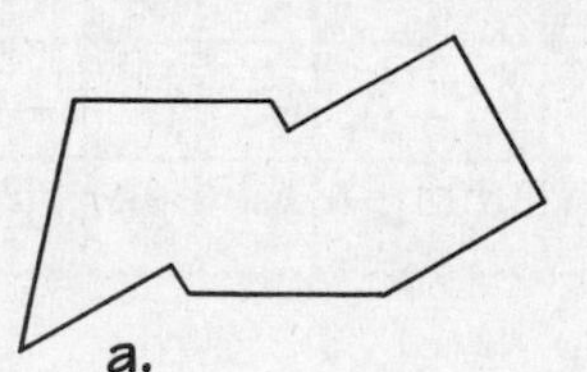

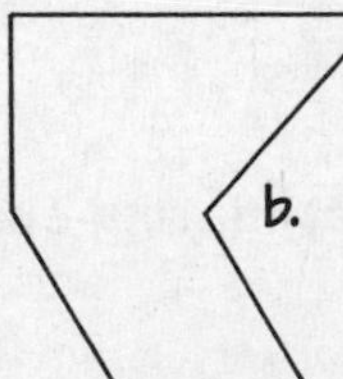

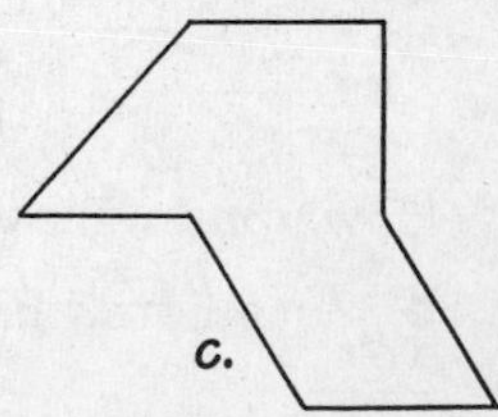

Check your answers against those listed on page 63. How well did you score?

3: Junior Detective

WHAT'S IN A NAME?

MATHEMATICAL OPERATION: Sequencing with patterns

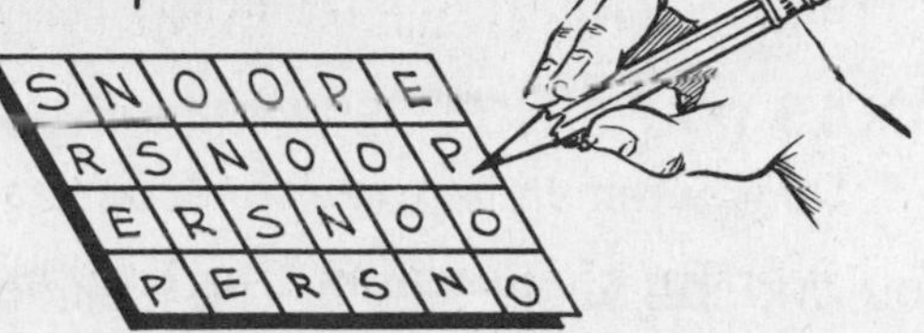

THE SCENE: Library, Super Sleuth Academy. You and the sleuths have fun with each others' names.

TOOLS YOU'LL NEED: Super Sleuth Journal, pencil, ruler

At a session at the Academy, the Super Sleuths are huddled over some mysterious squares. Pieces of paper lie on the floor with girls' and boys' names on them.

The Super Sleuths have a challenge. They have to find the patterns in names as they fill in various-sized grids. You can help. In your Super Sleuth Journal draw two boxes, one five squares by five squares, and the other, six squares by six squares. Write each letter of your first name in a square, moving from left to right. For example, the name Maria would look like this:

M	A	R	I	A
M	A	R	I	A
M	A	R	I	A
M	A	R	I	A
M	A	R	I	A

M	A	R	I	A	M
A	R	I	A	M	A
R	I	A	M	A	R
I	A	M	A	R	I
A	M	A	R	I	A
M	A	R	I	A	M

Lightly shade in every third square. Do you make a pattern? What kind of pattern do you make in the five-square box? In the six-square box?

PALS

MATHEMATICAL OPERATION: Palindromes

THE SCENE: Library, Super Sleuth Academy. You and the sleuths must calculate all the bad guys on record.

TOOLS YOU'LL NEED: Super Sleuth Journal, pencil

You and the Super Sleuths are looking through the book of *Suspects I Have Known* to find the mug shots of bad guys and good guys (undercover cops). You are to report to the Chief how many good guys there are and how many cheats, swindlers, and nasty crooks are in the book. How do you do it?

The sleuths are keeping a record in their journals of the criminals who have a palindrome as their "criminal number." A palindrome is a number that reads the same forward or backward. The key to finding the good guys can be found with palindromes, those numbers making them "pals," or good guys.

Here's what you do:

1. Write down a two- or three-digit number:

   Example: 34

2. Reverse the digits and add this number to the first one:

   Example: 43

   $$\begin{array}{r} 43 \\ +\ 34 \\ \hline 77 \end{array}$$

77 is a "pal": if you reversed 77, it is still 77!

See how easy it is? Now try it with the number 144. Take out your Super Sleuth Journal and a pencil and copy the number 144, reverse the digits, and add the numbers together.

144 + 441 = 585

The reverse is still 585. It's a palindrome!

If the sum is not a palindrome, keep reversing the number and adding it to itself until you come up with a "pal." If you don't get a palindrome until after you've reached three sums, you've found a "bad guy" who will never be good.

Here are the criminal numbers the Chief gave the sleuths to sort through. Help the sleuths pick out which ones are pals and which ones are crooks. List the good guys on the MOST UNWANTED list in your journal and the bad guys on your MOST WANTED list.

1. Criminal No. 994
2. Criminal No. 95
3. Criminal No. 82
4. Criminal No. 78
5. Criminal No. 918

DON'T BE SQUARE

MATHEMATICAL OPERATION: Matching patterns; deduction

THE SCENE: A dead-end alley. You and other sleuths must find a robber with the help of squares.

TOOLS YOU'LL NEED: Super Sleuth Journal, pencil

As the sleuths chase a robber down a street, he turns into an alley and . . . disappears! When the sleuths get to the alley, they find nothing more than the baseball cap the thief was wearing. But wait! Tucked into the brim of the hat is a tattered-looking note.

The sleuths find the note to be some sort of puzzle. It shows four different designs arranged into six different shapes. The sleuths have a hunch that this puzzle will disclose the crook's secret hiding place. Here are the four designs drawn on the paper:

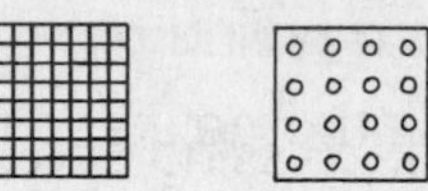

These squares have been configured (used in a special pattern) into the pictures below. Each pattern is different, but only one pattern has a certain specific difference from the others. This pattern is the one that will lead the sleuths to the crook. Can you help the sleuths find the right configuration? Fill in your answer in your journal.

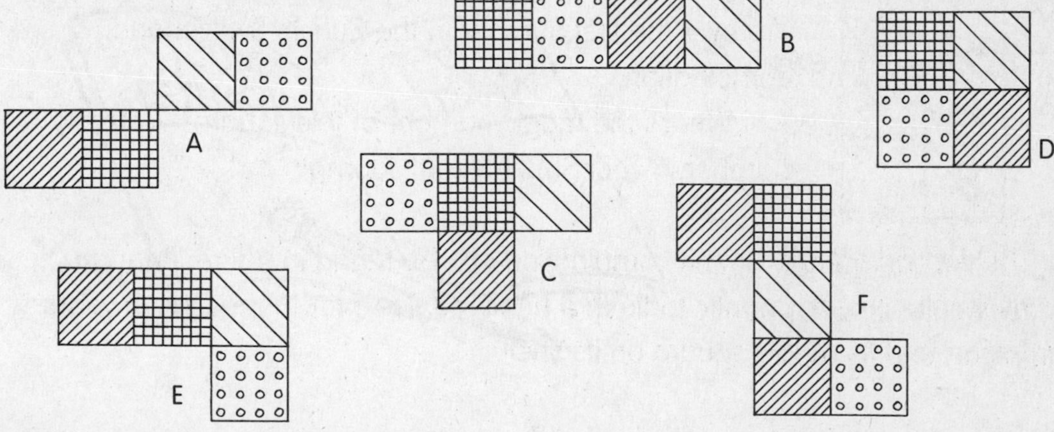

DESIGNS IN THE SAND

MATHEMATICAL OPERATION: Magic squares

THE SCENE: The Super Sleuths are flying to Israel, Germany, China, and Africa to solve the mystery of some strange designs that were left in a puzzle format called a magic square. If they find all four magic numbers, they'll have the secret combination to the safe deposit box in Zurich, Switzerland, where the name of the most deadly intelligence spy is hidden. You get to join them!

TOOLS YOU'LL NEED: Super Sleuth Journal, pencil

For your first clue, you and the sleuths must find the numbers that were missing in a magic sand square in Israel. A magic square is a square or design with numbers arranged so when you add up the numbers across, down, or diagonally, they always equal the same amount. Here's a sample square.

13	8	9
6		14
	12	7

Find the numbers in the missing spaces (you may have to try a few numbers before finding the right one).

Test its magic powers by adding across, down, and diagonally. If the total always equals 30, you've found the magic number of the square.

9		5
	6	10
	8	

Now try this magic square on your own to discover the first number in the Zurich, Switzerland, combination.

What's the magic number of this square? Write the number in your Super Sleuth Journal.

The second number to the combination can be found in China, near an old river called Lo. A gigantic turtle that has lived there for thousands of years carries around its magic square on its shell.

The Chinese painted circles on the turtle's back instead of writing numbers. Count them to deduce the correct number.

The magic number of this square is the sum of every row that intersects the middle circle. What is the missing number? The magic number? It will be the second number in the combination! Write down both the missing middle number and magic number in your journal.

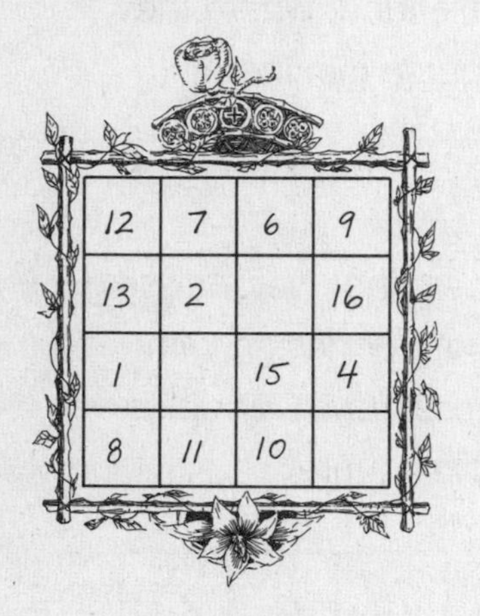

Next, you and the Super Sleuths must fly to Germany to uncover the third number in the combination. What numbers are missing in this magic square?

In your journal, write down its magic number (the sum of any row or column) as the third number in the combination.

Finally, you head to Africa, where they design their magic squares a little differently.

Here's how to solve it:

Step 1: Flip-flop the top and bottom middle squares: 3 and 7. Write them in the question-marked boxes in your journal.

		3		
	2		6	
1		5		9
	4		8	
		7		

=

	2	?	6	
	?	5	?	
	4	?	8	

Step 2: Flip-flop the left and right middle squares: 1 and 9. Fill in these numbers in the question-marked boxes as well. Now add up each row and column. You should get the same number every time, and that number is the fourth number in the secret combination!

Once you've helped the Super Sleuths find all four secret magic square numbers, you should know the combination to uncover the name of the secret intelligence officer.

What is it?

THE STAR SAPPHIRE CASE

MATHEMATICAL OPERATION: Logic and deduction

THE SCENE: The estate of the rich and famous Lord Brightly, Earl of Worth. A beautiful and fabulously expensive star sapphire has been stolen from the estate. The Super Sleuths have captured one robber and are hot on the trail of his accomplice.

TOOLS YOU'LL NEED: Super Sleuth Journal, pencil

You and the sleuths are following some bits of paper left across the carpet in the room where a hidden safe housed the missing star sapphire. The sleuths have captured one of the unscrupulous crooks, but the other remains at large. All clues lead to the time the crook's partner will return to try and free her. To help the sleuths, get out your Super Sleuth Journal and a pencil and copy the star sapphire pattern below.

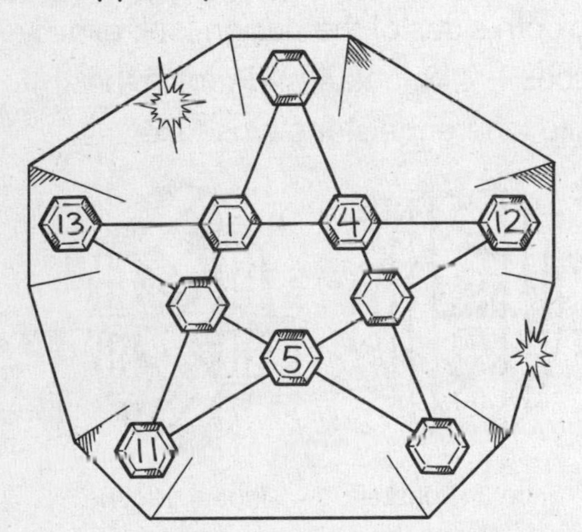

Each row—up, down, or diagonally—must add up to the same number. To find out what time the crook will try to spring his partner, read the clues below.

Clue 1: The five inner circles on the star sapphire are missing two numbers from 1 to 5. The sum of the two missing numbers is the hour the thief will return.

Clue 2: The five circles around the outside of the star sapphire are missing two numbers from 9 to 15. The sum of the two missing numbers is the exact minute the crook will come back.

The time for the rendezvous is ______:______ P.M.

SHIELDS

MATHEMATICAL OPERATION: Symmetry

THE SCENE: A small African village. You and the Super Sleuths must find a stolen ruby.

TOOLS YOU'LL NEED: Super Sleuth Journal, pencil

The Super Sleuths are questioning a suspect in Africa, where they have been sent on their assignment. A precious ruby has been stolen by one of the nearby tribes and must be returned so the annual Celebration of the Sacred Stone can occur.

It's up to you and the sleuths to fill in the puzzle to find which African tribe stole the fabulous ruby. A piece of a shield was found near where the stone was taken. Using your Super Sleuth Journal and your good "mind-sight"—using your mind and eyes—finish the design on the shield to make it symmetrical (if you split the shield exactly in half, both sides of the design look exactly the same). Compare the designs from tribes 1, 2, 3, and 4 and match the design on the piece of shield left at the site where the stone was taken.

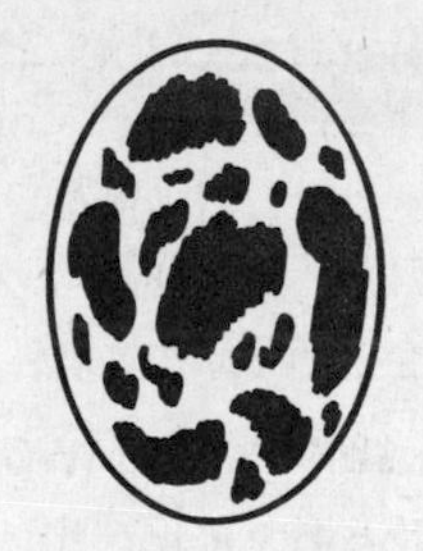

Tribe 1, Rwanda

Tribe 2, Kenya

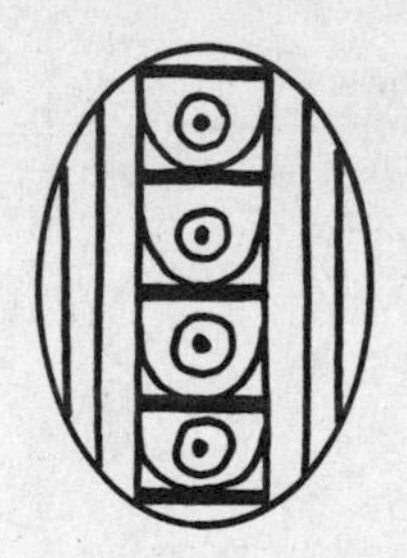

Tribe 3, Transkei

Tribe 4, Nigeria

This is the piece of shield that was found at the site:
Which tribe took the precious ruby?

MUG SHOTS

MATHEMATICAL OPERATION: Matching, probability

THE SCENE: Computer control room, police headquarters. You and the sleuths piece together the identity of an art museum thief.

TOOLS YOU'LL NEED: Super Sleuth Journal, pencil

The sleuths caught a quick look at a crook as he ran away from an art museum. At police headquarters, the computer-generated face-maker machine is ready and loaded up with eyes, noses, mouths, ears, and styles of hair. Can the sleuths rely on their memories, or have they forgotten what the culprit looks like?

The probability or the chance of finding what the crook looks like using the computer-generated face-maker machine is 1 in 243! To figure out the probability, take the clues and multiply them by each other. For instance, there are five different features to put together to make the criminal and three types of each feature. So, the equation would look like this: 3 (the number of eye types) x 3 (the number of nose types) x 3 (the number of ear types) x 3 (the number of mouth types) x 3 (the number of hair types) equals 243 possibilities, or different combinations.

It's up to you to pick the best descriptions using these clues:

1. Hair is ragged and spiky.
2. Ears are rounded and chubby.
3. Mouth turns down.
4. Nose is very crooked.
5. Eyes are droopy ovals.

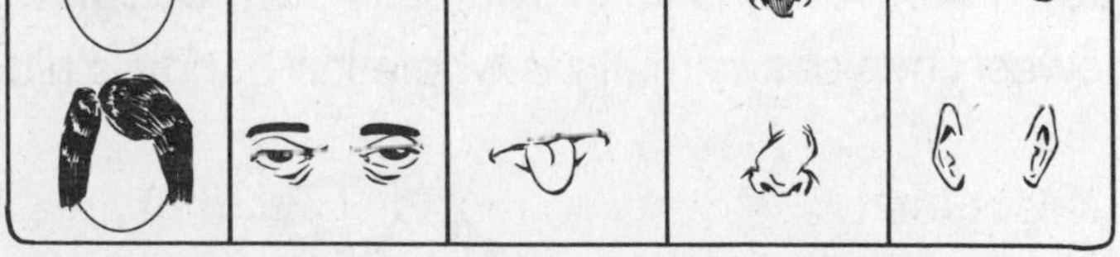

HAIR	EYES	MOUTHS	NOSES	EARS

Draw your best guesses in your journal and match them up with the computer printout for Public Enemy No. 11037.

MIRROR, MIRROR, ON THE WALL

MATHEMATICAL OPERATION: Symmetry

THE SCENE: A circus carnival near the Super Sleuth Academy. The proprietor of the House of Mirrors has called the sleuths to catch a quick-change artist.

TOOLS YOU'LL NEED: Super Sleuth Journal, pencil, small two-sided pocket mirror

The carnival's ticket earnings have been taken by a desperate bandit. He's dashed into the House of Mirrors, where many mirrors are lined up against the walls with pathways leading in and out between them. How can the sleuths tell if it's really the bandit, or just a mirror image of him? They have to know about the "line of symmetry" of a shape—the line along which a mirror is placed so that exactly one half of the shape is reflected in the mirror (the other half is blocked by the mirror). Think of the line of symmetry as finishing a half. Put the mirror along the center of Figure A to get a better idea of what a line of symmetry is and how you can make a mirror image.

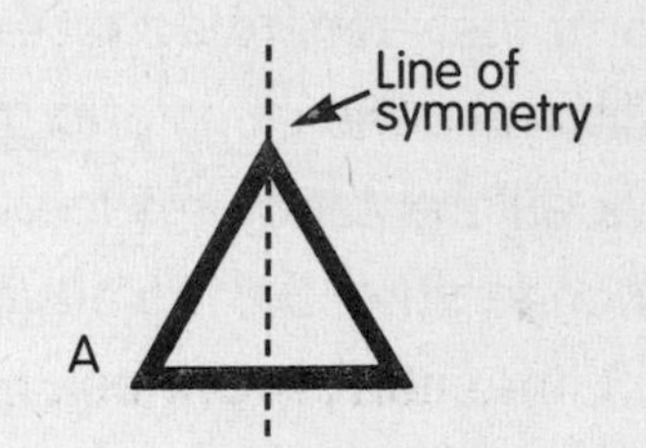

To find the bandit, take a two-sided mirror and put it along one of the image's line of symmetry. Remember, there may be more than one line of symmetry—it depends on the design. Move the mirror slowly around the shape, looking for other lines of symmetry. Every time the mirror makes the image a complete whole, you've found a line of symmetry. Next, in your journal, write down how many lines of symmetry each design has. The one that has the fewest lines of symmetry is where the bandit is hiding. Is he in figure B, C, or D?

B

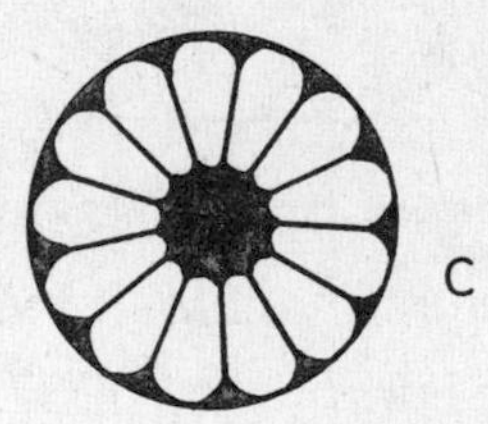
C

D

CHIP OFF THE OLD BLOCK

MATHEMATICAL OPERATION: Various operations

THE SCENE: A high-rise building in a neighborhood block. The Super Sleuths are chasing a notorious crook who's going in and out of different apartments with the secret plans for a revolutionary laser beam.

TOOLS YOU'LL NEED: Super Sleuth Journal, pencil, place markers

The Super Sleuths are hot on the trail to recover the valuable plans for the revolutionary laser beam. With his stolen copy of the building plans, the thief is able to dodge in and out of four floors of apartments following a special shortcut escape route he designed using his "Lucky 5" numbers. The trick is that he follows the path where an answer has a 5 in it; if there is no 5, he can't move from one apartment to another. For instance, he can move from 10 - 5 to 5 - 20, but can't go from 13 - 7 or 10 x 4.

You can help the sleuths find his escape route and where he plans to exit. Copy the block shown here in your journal or simply get pebbles, beans, or other markers to follow his tracks directly on the page. Exactly where will his exit be?

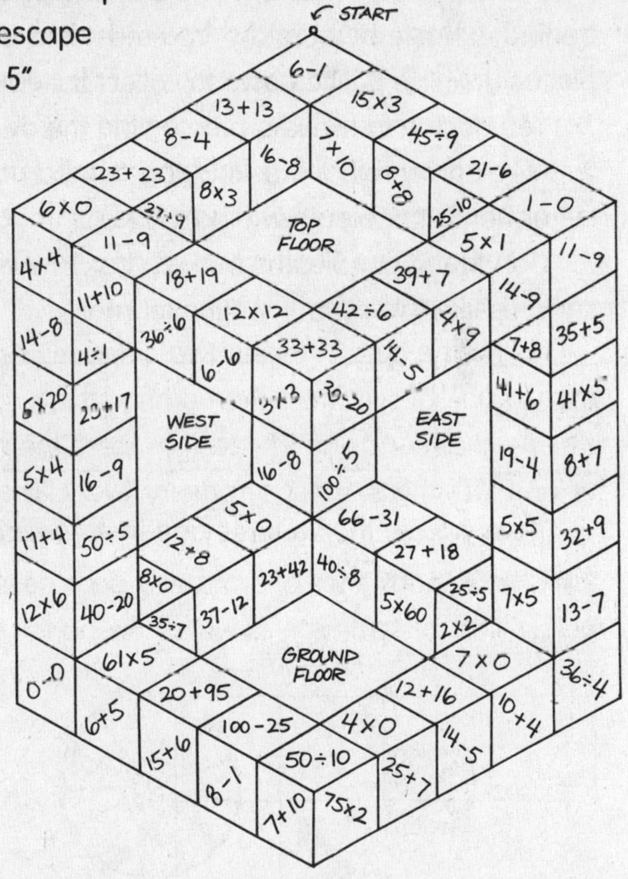

IT'S A WRAP-UP!

MATHEMATICAL OPERATION: Geometric design building

THE SCENE: An apartment building. You and five other sleuths must guard a diamond ring and necklace until police arrive.

TOOLS YOU'LL NEED: Super Sleuth Journal, pencil, scissors, masking tape

Outside an exclusive apartment building, late in the evening, you and four other Super Sleuths have found a sparkling diamond ring and necklace left behind in haste by burglars. You must find a safe place to keep these valuable pieces until the police arrive to collect the evidence.

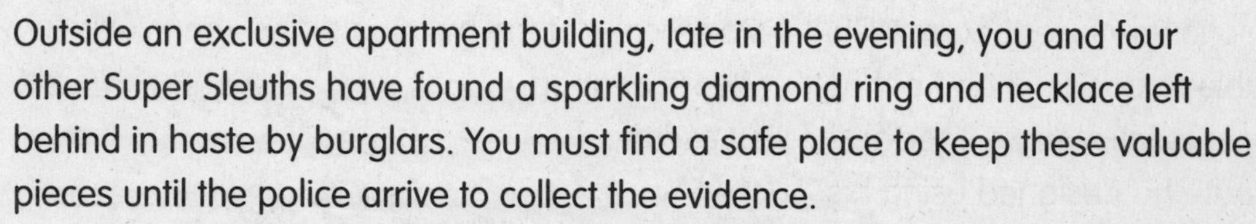

You decide to make a box to hold the evidence. But as you look around, you see only a roll of masking tape and a pair of scissors. Everyone does have his or her Super Sleuth writing journals.

Can the Super Sleuths make a box? Take out your pencil and draw five squares like this in your writing journal:

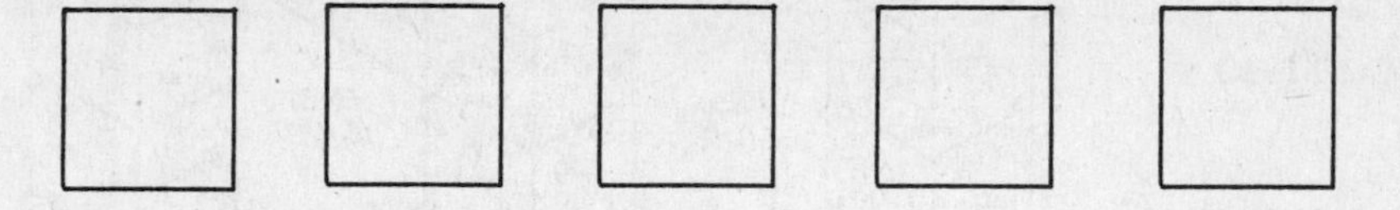

Then cut out the squares and try to arrange them so their edges touch and their corners line up . . .

like this:

but not like this:

Arrange the squares on a flat surface in such a way that, when the squares are folded, they form a box.

Here are a couple of clues:

Clue 1. There are twelve ways to arrange the squares.
Clue 2. There are four ways to make boxes.

Be careful! If one arrangement of squares fits exactly over another, only turned around, it doesn't count as a new arrangement. Here are two shapes that are the same, just turned around. They count just as one shape.

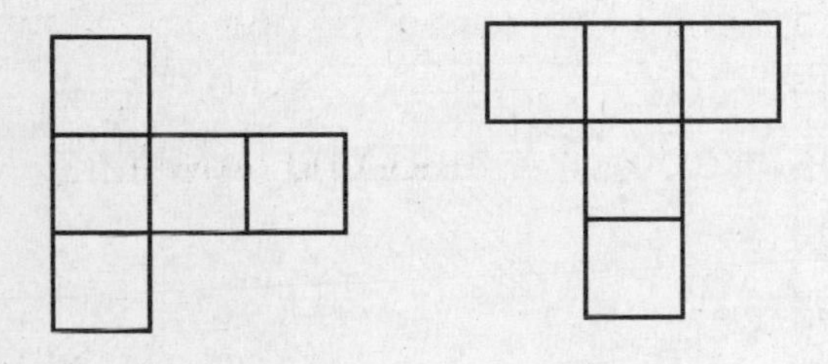

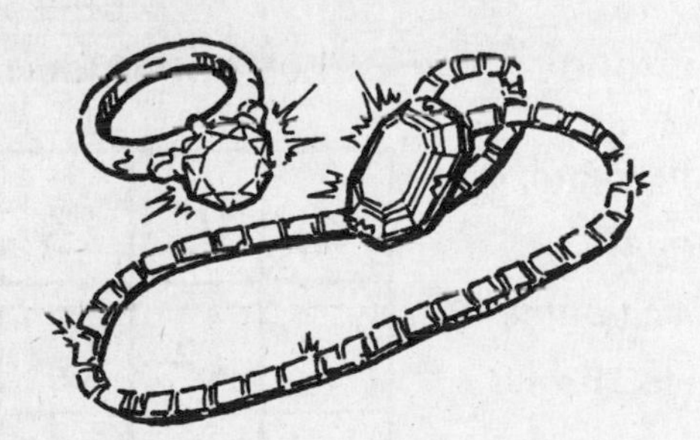

SUPER SLEUTH CHALLENGE

In this challenge, try using a different number of squares to fold into a box. Test your theory by folding the squares into a box, taping the edges of the squares together, and "boxing it." Stuff your boxes with presents for friends and family members!

JUNIOR DETECTIVE QUIZ

By this time your sleuthing skills should be getting razor sharp! Open your journal to a clean page to test your progress. Each quiz question corresponds with an earlier problem you faced in this chapter. If you get stumped, turn to the mystery noted for extra help.

1. Palindromes, or "pals," are those numbers that read the same way forward as they do backward. In "Pals," you added the criminal ID number by its reverse to see if the sum was a palindrome. If you came up with three sums and still had not reached a palindrome, the crook was indeed a bad guy. Check out these other numbers and see if they are pals or crooks.

   A. 916 B. 79 C. 488

2. In "Don't Be Square" you figured out which configuration didn't belong. Can you do it again?

   A. B. C. D.

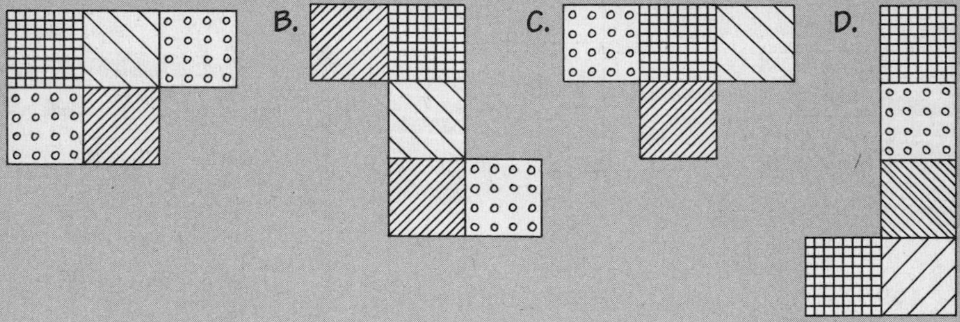

3. Using your skills from "Designs in the Sand," fill in the missing numbers of the square. Every row and column should equal the same amount (not diagonals). That total is the magic number. What is it?

<table>
<tr><td>9</td><td>14</td><td>13</td></tr>
<tr><td></td><td>12</td><td>8</td></tr>
<tr><td>11</td><td></td><td></td></tr>
</table>

4. In "The Star Sapphire Case" you sharpened your skills in number patterns. What are the missing numbers and what is the magic number?

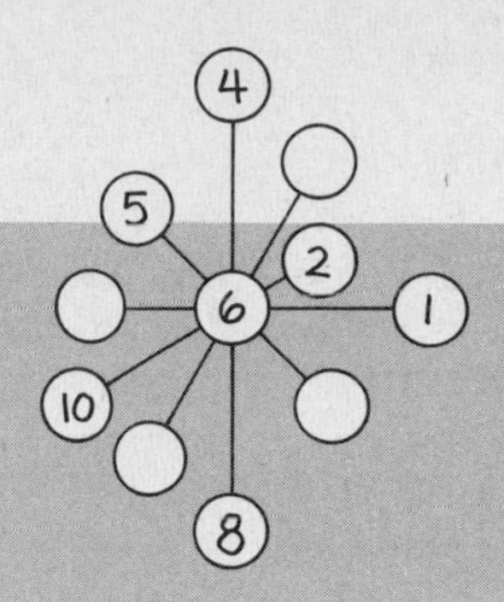

5. With a ruler, construct a seven-box grid and an eight-box grid. Beginning in the top left corner, write your last name in the grid from "What's in a Name?" Shade in all the vowels. Does it make it a pattern? What is the difference between the seven-box pattern and the eight-box pattern?

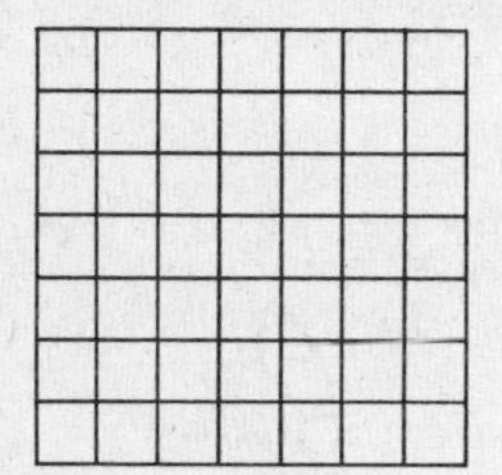

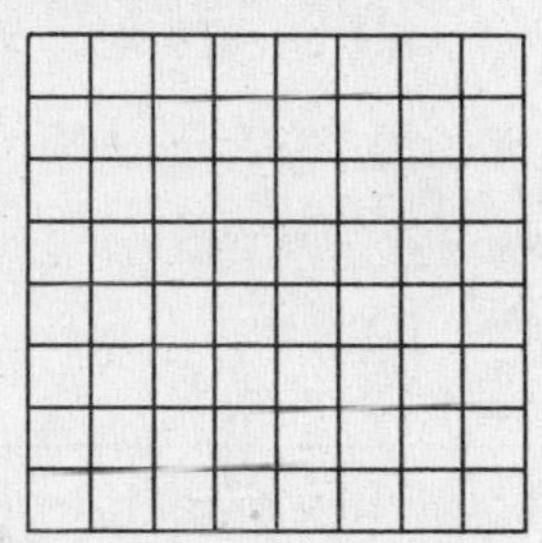

START ↓

75x2	100−25	23+42	3x3	8+8
17+4	16−9	37−12	16−8	39+7
12x6	40−20	66−31	19−4	4x9
0−0	33+33	6−6	8+7	42÷6
14−8	27÷9	11+1	15x3	6−1
4÷4	8−4	36÷6	42÷6	7x5
	13−6	12x12	3x13	20+95

6. As you did in "Chip off the Old Block," to get out of the maze, follow the Lucky 5 numbers.

7. Turn back to "Mug Shots" and take a look at the facts on probability. How many combinations can you make if each feature has two possible designs?

Eyes | Hair | Nose | Mouth

Check your answers against those listed on page 64. How well did you score?

SUPER SLEUTH WHIZ QUIZ

You've worked hard and had a lot of fun in ***Super Math Tricks****. Now let's see how well you've learned your detective lessons.*

Here's your assignment: A gang of dog-nappers has taken Sniffer, your K-9 buddy. To find him, you must get through this weird and puzzling maze and record all your answers in your journal. To stay clear of trouble, you'll have to do your sleuthing with great care . . .

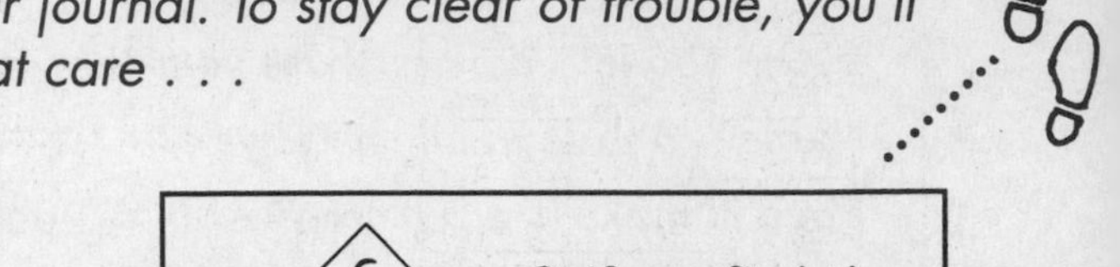

1. Is this a palindrome? If it is, go on.

1 5 5 5 1

2. Is the answer a palindrome? If yes, go on.

$$\begin{array}{r} 10010 \\ -\ \ 9999 \\ \hline \end{array}$$

3. Find what the next object should be.

4. Find what the missing number is, then add up each line to discover the magic number.

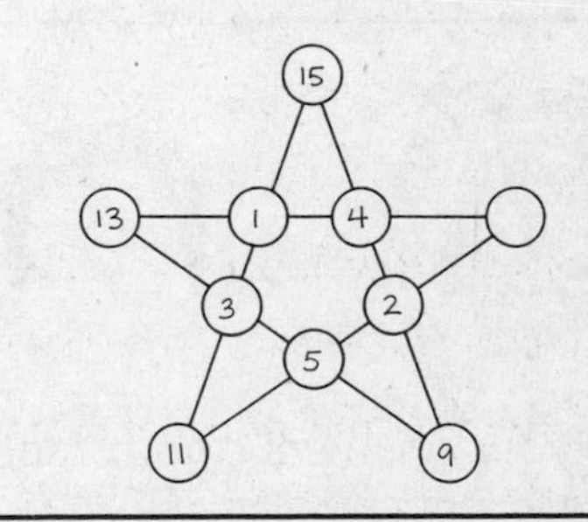

5. What should you do to get over the log? Use the key below to solve the numeric code.

U=1 P= 2 M=3 J=4

CODE: 4-1-3-2

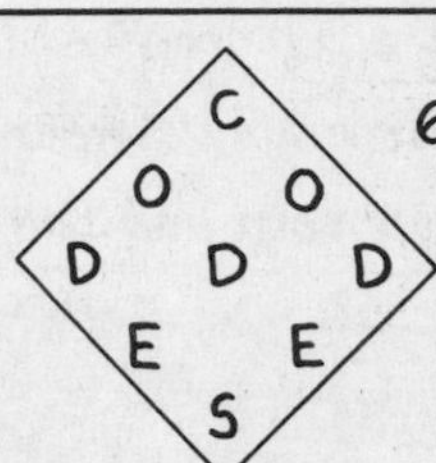

6. If you find the word in the diamond, go on.

7. To get over the iceberg, name its shape, then proceed ahead.

START ↓

15−10	20+5	30×5
27+12	6×0	7×5
23-23	3×3	6 −1
4÷1	21-6	45÷9
14-9	5×1	17+4
41×5	22×2	6 −6
37-12	27+18	5×5
8×0	13+13	66-31

8. To pass through the mine field, you must have a Lucky 5 as an answer. Can you make it through?

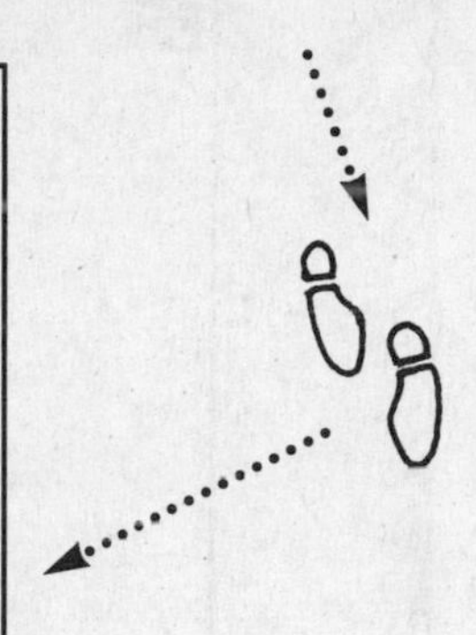

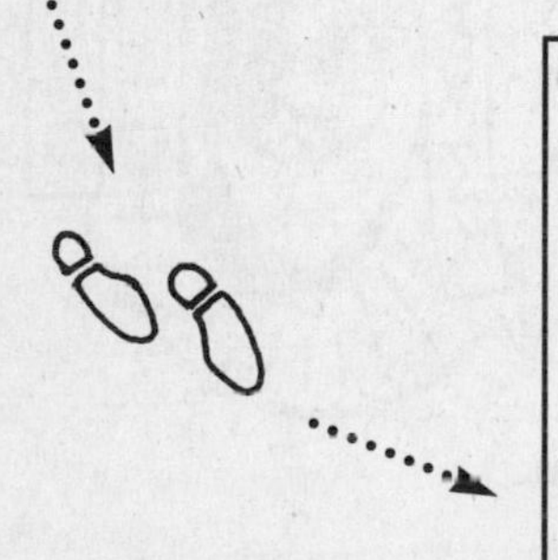

9. Using the key below, can you buy these words (CHASE ME) if you have three quarters? If you can, go on.

Key:

A = 5 cents C = 9 cents S = 17 cents
E = 13 cents H = 10 cents M= 6 cents

10. Check your answers on pg. 64. If you got them all right, you did it! Put Sniffer back in his doghouse. Good work, and welcome to the Super Sleuth Detective Force! (If you missed some questions, go back and see what you did wrong.)

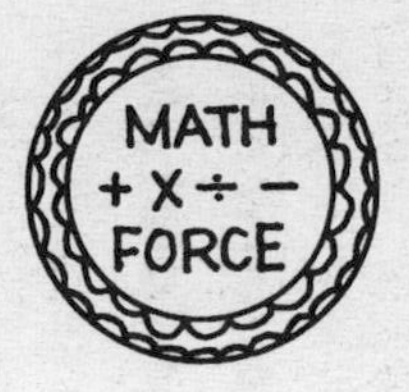

Certificate of Completion

SUPER SLEUTH TRAINING ACADEMY

NAME: ______________________ DATE: ______________

Blaise Pascal
HEAD MATHEMATICIAN

René Descartes
CHIEF OF MATH

Answers

CHAPTER 1

Get the Facts p. 6 *The numbers on the left should be your telephone number. The number (or numbers) on the right makes up your age.*

Badge Me p. 8 *Message: IF YOU CAN SOLVE THIS, YOU DESERVE A BADGE!*

The Case of the Secret Message p. 10 *Message: SUPER SLEUTH MATH TRICKS ARE COOL. FIGURE THEM OUT, DON'T BE A FOOL!*

Agent G.'s Mysterious Code p. 11 *Message 1: 789-4326; Message 2: 789-4326; Message 3: 789-4325. Secret Agent G's telephone number is the number that is different: 789-4325.*

Repeat-a-Crime Code p. 13

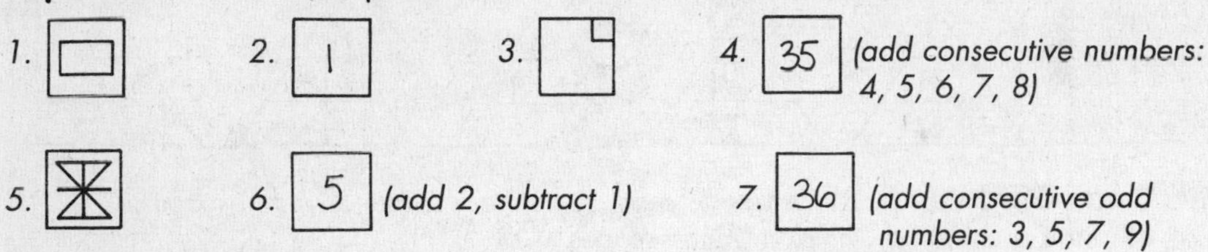

How Do You Measure Up? p. 15 *1. foot:neck ratio—Answers will vary, such as: If the length of your foot equals the length around your neck, the ratio should be 1:1.*

2. neck:wrist ratio—Answers will vary, such as: If the length of your neck equals the length around your wrist two times, the ratio should be 1:2.

3. elbow to longest finger:neck ratio—Answers will vary, such as: If the length from your elbow to your longest finger equals two lengths around your neck, the ratio should be 1:2.

Find the "Funny Money" p. 17 *1. blocks 11, 22, 33, 44, 55, 66, 77, 88, 99 are filled in 2. 5, 15, 25, 35, 45, 50, 51, 52, 53, 54, 55, 56, 57, 58, 59, 65, 75, 85, 95 are filled in 3. 20, 31, 42, 53, 64, 75, 86, 97 are filled in 4. 9, 18, 27, 36, 45, 54, 63, 72, 81, 90 are filled in 5. 7, 14, 21, 28, 35, 42, 49, 56, 63, 70, 77, 84, 91, 98 are filled in Block 91 is where the counterfeiters are hiding.*

Hot or Cold? p. 19 *1. Kona, Hawaii 2. Fairbanks, Alaska 3. Seattle, Washington 4. Los Angeles, California 5. Kona and Los Angeles 6. Since she's a surfer girl who loves palm trees and drinks tons of guava punch, she'll probably be in Kona, Hawaii.*

The Right Combination p. 20 *Column 1: 29723 Column 2: 23846 Column 3: 23484*

The Stepping-Stone Number Chase p. 22 *1. 10; hop over 2. 12; hop over 3. 25; step 4. 11; step 5. 10; hop over 6. 10; hop over 7. 15; step 8. 21; step 9. 24; hop over 10. 1440; hop over*

Math Sleuth Rookie Quiz p. 24 *1. your address 2. GREAT WORK 3. 125-3491 4. 21 (add consecutive numbers: 1, 2, 3, 4, 5, 6) 5. I WANT YOU TO REACH THE TOP OF YOUR CLASS! 6. Answers may vary 7. 8 8. a) 14 b) 12 c) 21st d) 11; common number is 1*

CHAPTER 2

The Cutup Caper p. 26 *A is Shape 3; B is Shape 1; C is Shape 4; D is Shape 2.*

The Pyramid's Chain p. 28

Message: I N A C A M E L ' S H U M P
27 23 19 21 19 20 22 34 30 10 18 16 14 24

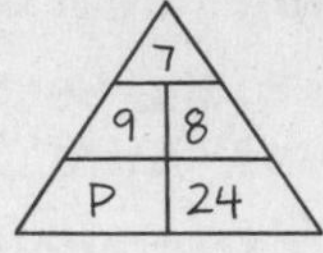

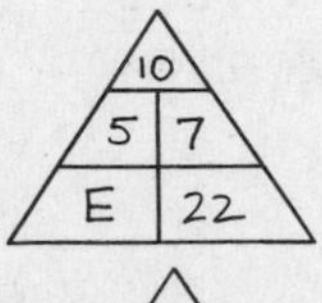

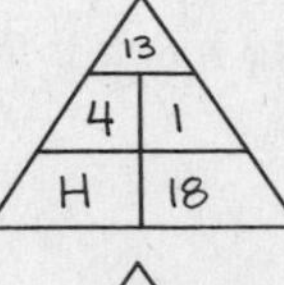

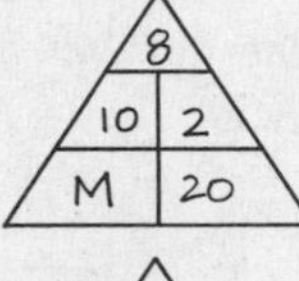

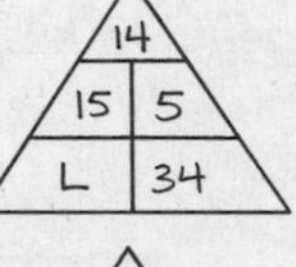

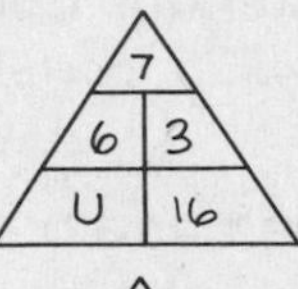

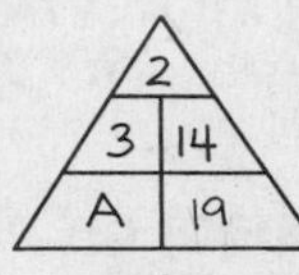

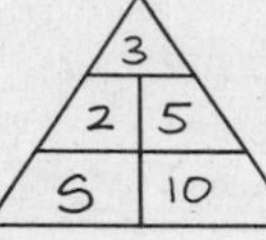

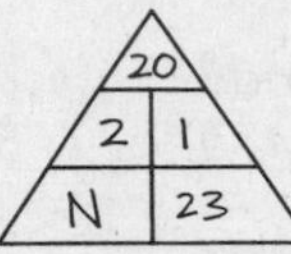

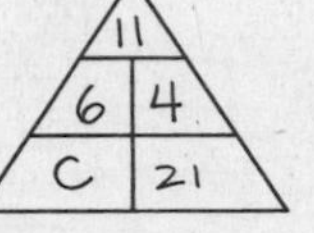

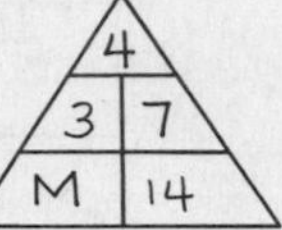

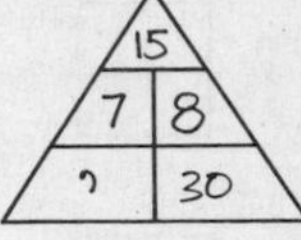

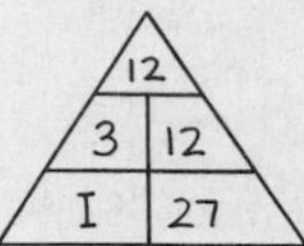

The Clean Getaway p. 29 *Maze 1: man is captured; Maze 2: man escapes; Maze 3: man is captured; Maze 4: man escapes*

Shape It Up p. 31 *The message is in Circle 1. It says: DIAMONDS ARE UNDER THE CAR SEAT.*

Memory Madness p. 33 *Answers will vary.*

The Vanishing Crook p. 34 *When you slide the lower piece of paper to the left, one of the faces disappears. You now have five faces instead of six.*

Tangram Twins p. 35 *The eighteen pieces fit into the shape of a cat; the bag with one million dollars is in the cat box. (See figure at right.)*

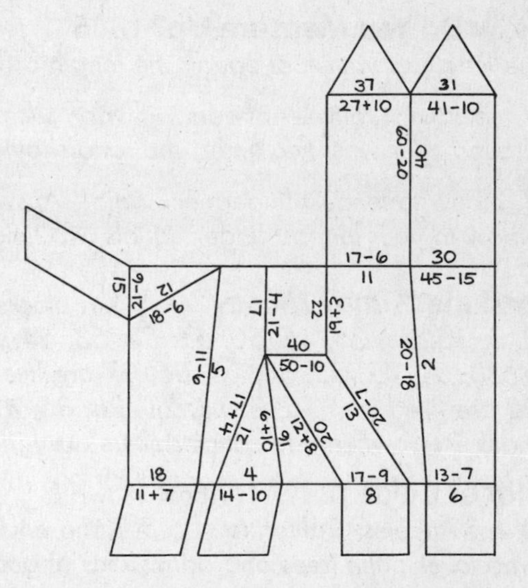

More Mysterious Shapes p. 37 *The answer is Figure 3 (below).*

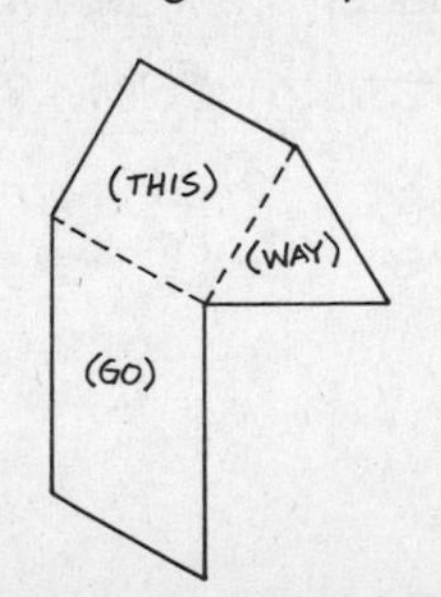

The Fractured Fraction Ferris Wheel p. 38 *The empty seat is No. 16.*

Waistline Bandits p. 39 *Belt 1: You get two separate belts that look the same—Jim and his look-alike twin, John, robbed the toy section.*

Belt 2: The belt has doubled in size—big enough for Fat Fred, the obese thief, who stole from the jewelry department.

Belt 3: The two belts are intertwined—inseparable Sue and Sally took several pairs of shoes from footwear.

Math Sleuth Cadet Quiz p. 41 *1. IN A CUP* *2. Circle 1 says YOU'RE A WINNER*
3. Answers will vary
4. The loot is in an aquarium. *5. a) $\frac{1}{4}$ b) $\frac{1}{3}$ c) $\frac{3}{6}$* *6. a*

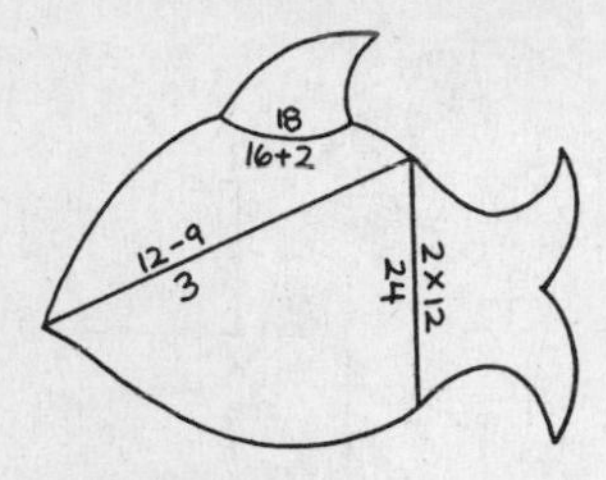

7. c; IN A BOX

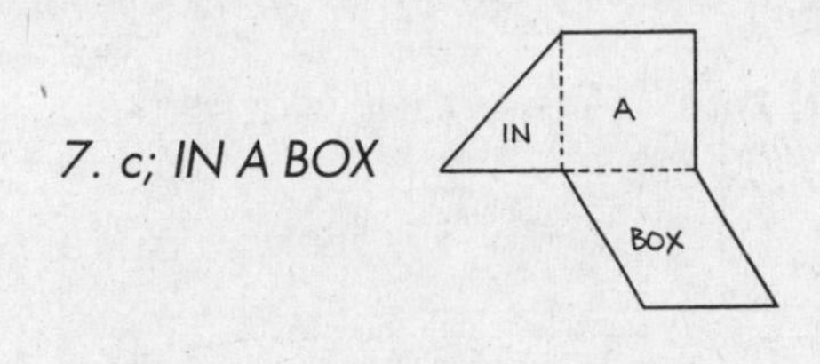

CHAPTER 3

What's in a Name? p. 43 *When every third letter is shaded in, the five-unit grid makes a diagonal pattern. In the six-unit grid, columns are made.*

Pals p. 44

Most Unwanted	Most Wanted
1. 9779 yes	*4. 4884 no—takes four times to reach palindrome*
2. 1111 yes	*5. 89298 no—takes four times to reach palindrome*
3. 121 yes	

Don't Be Square p. 46 *All four patterns are used in groups A, B, C, D, E, and F. Their configurations—the way they are placed in space—are different. Only group F has five squares, and two of them have the same pattern—they're diagonals.*

Designs in the Sand p. 47 *sample square: row 2, 10; row 3, 11*
first magic square: row 1, 4; row 2, 2; row 3, 7, 3
second magic square, River Lo turtle: center circle, 5
third magic square, German square: row 2, 3; row 3, 14; row 4, 5
fourth magic square, African square: row 1, 7; row 2, 9; row 3, 3
The combination to the safety deposit box is: 18 - 15 - 34 - 15

The Star Sapphire Case p. 49 *The missing numbers in the inner circles are 3 and 2; the missing numbers in the outer circle are 15 and 9. The time for their rendezvous must be 5:24 p.m.*

Shields p. 50 *The sleuths will find the ruby in Tribe 3, the Transkei Tribe village.*

Mug Shots p. 51 *Printout for Public Enemy No. 11037 (right)*

Mirror, Mirror, on the Wall p. 52 *Figure B, 4 lines; Figure C, 13 lines; Figure D, 2 lines (below). The bandit is hiding in Figure D.*

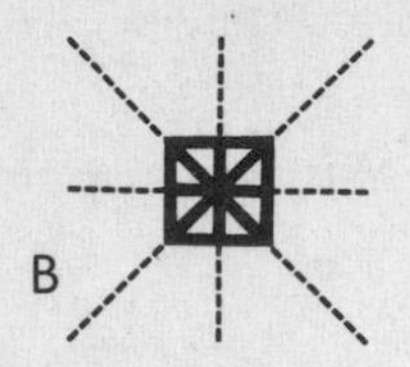

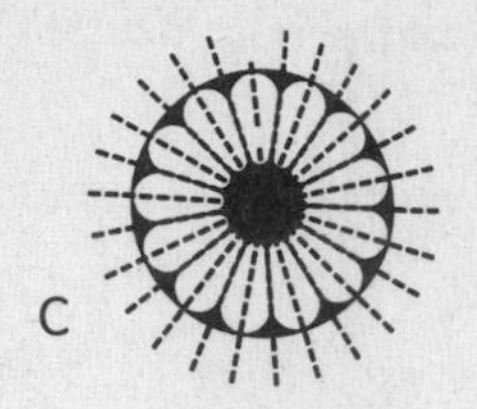

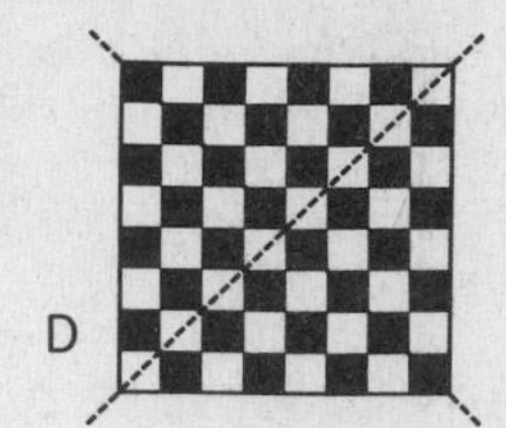

Chip off the Old Block p. 53 *He'll exit from a corner apartment on the ground floor. (See figure at right.)*

It's a Wrap-Up! p. 54 *Answers will vary (see below for examples).*

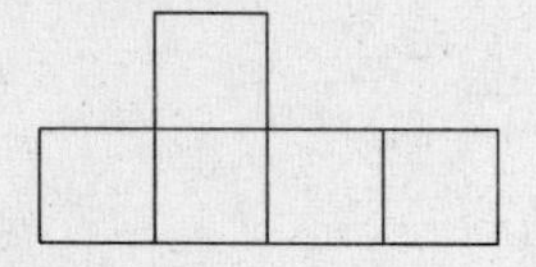

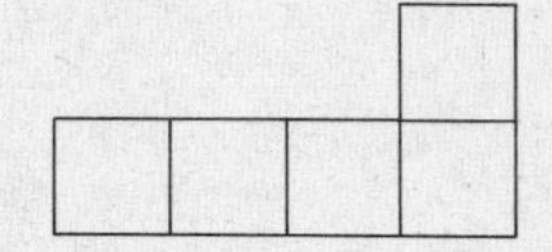

Junior Detective Quiz p. 56

1. a: yes, 6886 b: no, it takes six times to reach 44044 c: yes, 7117

2. C (has only 4 squares)

3. row 2, 16; row 3, 10, 15; magic number is 36

4. clockwise from the top: 3,7,9,11; magic number is 18

5. answers vary

6.

START 75x2	100-25	23+42	3x3	8+8
17+4	16-9	37-12	16-8	39+7
12x6	40-20	66-31	19-4	4x9
0-0	33+33	6-6	8+7	42÷6
14-8	27÷9	11+1	15x3	6-1
4÷4	8-4	36÷6	42÷6	7x5
	13-6	12x12	3x13	20+95

7. 16 possible combinations (2 sets of eyes x 2 types of hair x 2 noses x 2 mouths)

Super Sleuth Whiz Quiz p. 58 *1. yes 2. yes, 11 3. star 4. missing number is 12; magic number is 30 5. JUMP 6. CODES 7. triangle*

8.

START 15-10	20+5	30x5
27+12	6x0	7x5
23-23	3x3	6-1
4÷1	21-6	45÷9
14-9	5x1	17+4
41x5	22x2	6-6
37-12	27+18	5x5
8x0	13+13	66-31

9. yes, it's 73 cents